Infrared Spectra of Minerals and Related Inorganic Compounds

To my family,
especially my wife Brenda

Infrared Spectra of Minerals and Related Inorganic Compounds

J. A. GADSDEN ARIC

BUTTERWORTHS

BUTTERWORTH GROUP ADDRESSES

THE BUTTERWORTH GROUP

ENGLAND
Butterworth & Co (Publishers) Ltd
London: 88 Kingsway, WC2B 6AB

AUSTRALIA
Butterworths Pty Ltd
Sydney: 586 Pacific Highway, NSW 2067
Melbourne: 343 Little Collins Street, 3000
Brisbane: 240 Queen Street, 4000

CANADA
Butterworth & Co (Canada) Ltd
Toronto: 2265 Midland Avenue,
Scarborough, Ontario, M1P 4S1

NEW ZEALAND
Butterworths of New Zealand Ltd
Wellington: 26–28 Waring Taylor Street, 1

SOUTH AFRICA
Butterworth & Co (South Africa) (Pty) Ltd
Durban: 152–154 Gale Street

USA
Butterworth
161 Ash Street,
Reading, Mass. 01867

First published 1975

ISBN 0 408 70665 1

Typeset by Amos Typesetters, Hockley, Essex

Printed and bound by R. J. Acford Ltd,
Industrial Estate, Chichester, Sussex

Preface

The importance of infrared spectroscopy as a complementary technique to X-ray diffraction in the identification of minerals has become increasingly appreciated during the last decade or so. The infrared analytical technique is extremely sensitive to short-range site occupancy differences and thus may be contrasted with X-ray diffraction techniques which rely mainly on longer range ordering and on the periodic repetition of atoms. As with X-ray diffraction, infrared spectroscopy can give a 'fingerprint' of a mineral that is uniquely characteristic; however, a collection of standard reference spectra covering the wide range of minerals is not available.

In theory, the spectroscopist cannot do better than prepare his own spectral collection by recording standards on his own instrument, but this is expensive, time consuming and is open to the criticism that for the spectra to be really valid an analysed sample should be used, and not a sample from a specimen, another part of which has been analysed.

This book is intended as a reference for those working in the field of mineral and inorganic identification and whilst not being a comprehensive compilation, does contain infrared absorption band tables for over 600 minerals and some 100 other related inorganic compounds. References are also included to a further 550 minerals and inorganic compounds.

In addition to the collection of band tables and the associated bibliography, sections are included dealing with sample handling, spectra-structure and group frequency correlations. With some practice, the use of this book should enable the spectroscopist to narrow his field of search down to a few likely candidates, which are referenced to the literature wherein the original spectra are to be found. Confirmation by direct comparison with a standard should complete the identification. It is hoped that this book will help to overcome the lack of reference data, perhaps the major impediment to the fuller use of this highly promising complementary technique to X-ray diffraction in mineral analysis identification.

1975 J. A. GADSDEN

Contents

Acknowledgements

The data were collected during the course of the past ten years in the Physical and Chemical Services Division of the Warren Spring Laboratory. The author could not have achieved the collection of the relevant data from the voluminous records without the help of Miss J. Holroyd and Messrs. Gillham and Winter of the Laboratory. The completed manuscript was read and helpfully criticised by Dr. I. Dunstan, Deputy Director of the Laboratory. Special thanks are accorded to Mrs. L. Gadsden who assisted in the preparation of the typescript.

The author's thanks are extended to the Director, Warren Spring Laboratory, Stevenage, for permission to publish this book.

PART I

Inorganic Infrared Spectroscopy

INTRODUCTION

The application of infrared spectroscopy to the identification of unknown organic compounds is well established and widely used. Infrared spectra of organic compounds generally exhibit sharp well-defined bands, assignable to the vibrations of the individual functional groups of which the molecule is comprised. However, the infrared spectra derived from inorganic compounds usually have absorption bands that are broad and overlapping, making assignments and the specific identification of a cation–anion pair more difficult.

Early applications of infrared spectroscopy to the study of inorganic compounds proved the usefulness of the technique for the identification of polyatomic anions of the type XO_4^{2-}, XO_3^-, etc. However, the range of the spectrometers was limited (4000–650 cm^{-1}) and a vast number of inorganic vibrations which occur in the 'far-infrared' region (e.g. interatomic vibrations in binary compounds, borides, oxides, sulphides, etc.) remained undetected. Modern infrared spectrometers extend the range to 250 cm^{-1} and beyond. The availability of this new instrumentation, coupled with the advance of co-ordination chemistry, has awakened a new interest in the application of infrared spectroscopy to inorganic compounds.

Although advocated as an aid to the identification of minerals by some workers, e.g. Lyon[1,2], Omori[3,21], Achmanova[5], Moenke[9,10], Alexanian *et al.*[585], etc., the introduction of infrared spectroscopy into the field of mineral analysis has been limited due to the excellence of the X-ray diffraction technique. However, the infrared technique is quick, simple, relatively inexpensive and serves as a useful supplementary technique in the identification of mineral species.

The use of infrared spectroscopy in the analysis of lunar materials has been considered by several workers[521], including Estep[523] and Perry[524], and recently the application of the near-infrared and ultraviolet to remote sensing has been studied by Hunt and Salisbury[542,543] and Hunt, Salisbury and Lenhoff[578].

As with the infrared identification of organic compounds, it is essential, before any reasonable claim to identification of a mineral species is made, to compare the unknown with standard reference spectra. The spectroscopist can do no better than prepare, from hand-picked specimens, his own spectral collection on his own instrument. This is, however, both laborious and time-consuming and still open to criticism unless the reference mineral is known to be a well-characterised specimen and not just a sample from an analysed specimen. A further complication arises from the sensitivity of the technique. Only small amounts of material are examined (0.1–10 mg depending upon the method of sample preparation) and extreme care in the selection of both sample and reference must be exercised to ensure authenticity.

The alternative is to rely on spectra published by other workers. Experience shows that location of these spectra in the literature is not always easy and is often time-consuming and fruitless. The books by Greenwood, Ross and Straughan[473,507] have helped in this respect.

It is the aim of the present book to give the mineral/inorganic spectroscopist a short cut to the information he requires. Later sections deal with sample selection and preparation, and interpretation of the resultant spectrum. Part III contains the spectral data of minerals (natural and synthetic) and other related inorganic compounds, collected from the literature, tabulated together with approximate relative intensities and some band shape information. The references to the original work are given in Part IV. Since the data have been collected from the literature, the mineral classification, purity and origin quoted by the authors is relied upon, but some assessment of quality has been made. A detailed discussion of the theory of infrared spectroscopy is not considered necessary in the context of this book, as this has been dealt with elsewhere[8,24–26].

Other collections of inorganic spectra include the early work of Miller and Wilkins[7] and, more recently, of Nyquist and Kagel[4] which contains almost 900 spectra, recorded in the 3800–35 cm^{-1} region, but of which less than 25 are named minerals. The Sadtler[6] collection and the book by Nakamoto[8] are biased towards co-ordination compounds. Those with mineralogical bias will find the works of Omori[3,21], Lyon[1] and Moenke[9,10] of most use. Omori reports 93 minerals and Lyon deals especially with the silicate minerals. The two major collections by Moenke, covering 473 minerals, are probably the best seen to date. The recent work of Lazarev[500] deals in considerable detail with the vibration spectra of silicates but only a handful of spectra are included.

APPLICATIONS OF INFRARED SPECTROSCOPY TO MINERAL STUDIES

The frequencies at which a material absorbs infrared energy depends upon the internal vibrations of the molecule and hence upon its composition. Polyatomic anions of the type SO_4^{2-}, NO_3^-, CO_3^{2-}, etc., absorb infrared energy of different wavelengths, and can be distinguished one from another. An empirical approach to assignment has been adopted and some success in stating the general nature of the unknown is now possible, without recourse to reference spectra, provided that the data are interpreted with caution. As the position of the maximum absorption in the infrared spectrum is sensitive to changes in mass and charge of the constituent ions, the experienced spectroscopist can place a species within a fairly narrow composition range.

There are four main areas in which infrared spectroscopy can be useful in mineral analysis:

1 Identification of minerals of different (but constant) chemical composition.
2 Minerals of constant composition but with different crystal structures, e.g. SiO_2 as quartz, cristobalite or tridymite, can be recognised.
3 The technique can allow distinction between minerals that exhibit a range of composition within a group.
4 The technique can be used to distinguish between minerals that vary in both structural modification and chemical composition.

However, one cannot overemphasise the caution with which the spectroscopist must approach mineral analysis. The technique is extremely sensitive to certain common minerals, e.g. calcite or α-quartz, and small percentages of these, as impurities, can produce marked variations in the resultant spectra. Care in sample selection is vital. The purity of both the sample and the reference material must be beyond doubt. Also prevalent are ion-exchange effects between the alkali metal halides, used in the pelleting technique and as supporting plates for mulls, and the cations in the sample.

It must be borne in mind that comparison with the spectra published in the literature relies on reproductive accuracy as much as those assumptions as to purity. It is wise to accept the authors' quoted absorption maxima (assuming adequate instrument calibration) rather than rely solely on a visual examination, although the latter is naturally useful as band contours are important.

Infrared spectroscopy is especially useful in the investigation of short-range ordering within a mineral group, and becomes of paramount importance when dealing with 'amorphous' material for which X-ray crystallographic data are tenuous; see, for example[508], Ptak and Pampuch[536], marine minerals by Elderfield[613] and the various works with glasses[198,673], etc. That there is considerable potential for the use of infrared spectroscopy in structural studies of minerals has been pointed out by Adler[669] and Solntseva[535], and expanded more recently by Lazarev[500] and Povarennykh[621] among others.

The following examples serve to illustrate the areas in which studies have been made recently:

(*i*) The structure of the OH group in phlogopite and biotite micas, by Plyusnina and Kapitnova[658] and the orientation of OH in minerals, by Zemann[663].
(*ii*) A study into the degree of ordering in K and Na feldspars, by Kuznetsova[659].
(*iii*) Isostructural minerals have been the subject of studies by Povarennykh alone[660] and with Gevorkyan[661].
(*iv*) The nature of bonding and oxygen–halogen co-ordination of cations in minerals, by Solntseva *et al.*[662].

A review of the advantages and disadvantages of infrared spectroscopy in the study of mineral isomorphism (with 86 references) is given by Akhmanova and Alekhina[665].

It can be seen that, if the above-mentioned difficulties are borne in mind, infrared spectroscopy can serve as a very useful supplementary technique to X-ray diffraction in mineral investigations.

INSTRUMENTATION AND EXPERIMENTAL TECHNIQUES

The spectrometer

The region covered by the older prism spectrometers (5000–650 cm^{-1}) does not extend far enough into the infrared to cover many of the important inorganic molecular and lattice vibrations, neither do they, generally, have adequate resolution in the N—H and O—H stretching regions. A modern grating spectrometer extending the range to at least 400 cm^{-1} is required, and can be expected to have the necessary resolving power. Although instrumentation becomes more expensive as the range is extended, it is worth remembering that information from frequencies below 400 cm^{-1} is becoming increasingly available. No particular emphasis is given to far-infrared work in this book, but reference to papers giving far-infrared spectra is made where it is relevant. A useful assessment of some of the instruments now available is given by Ferraro[12].

Calibration of the spectrometer must be carried out at regular intervals, to ensure accuracy in the comparison of unknowns with reference spectra. A number of calibration points are required and those quoted in the I.U.P.A.C. *Tables of Wavenumbers for the Calibration of Infrared Spectrometers*[13] are recommended.

Sampling techniques

Details of sample preparation procedures are given in other texts (Stanfield, Shepperd and Harrison[14], Stewart[184]) so discussion here will be restricted to those aspects of the procedures that are pertinent to inorganic/mineral analysis.

SAMPLE SELECTION

It is essential that the material presented to the spectrometer is composed solely of the species to be analysed. In this respect minerals can present something of a problem, as in the natural state they often occur as intimate mixtures of several different species. Such mixtures may be unconsolidated (e.g. sands and soils) or consolidated as in the case of sandstones and granites. Prior to isolating a monomineralic fraction for analysis, consolidated material must be crushed to a sufficiently fine size to separate the various contained minerals from each other.

Several techniques are available for the separation of the required mineral from the bulk. As only a small sample is needed (from a few grains to 10 mg), hand picking with the aid of a microscope is recommended, but accuracy in selection is essential. However, identification and separation of the mineral becomes increasingly difficult as the particle size is reduced, and other techniques must be applied, e.g. magnetic and density separations. In addition, selective chemical attack, e.g. acid dissolution, may be useful to remove traces of contaminants. Laboratory methods of mineral separation have been reviewed by Muller[27]. It is essential that the separated mineral be cleaned of the organic liquids used in the separation techniques, prior to recording the infrared spectrum.

Further grinding of the selected sample, to a particle size below the wavelength of the radiation being used, is necessary if loss of energy from the infrared spectrum, by scattering, is to be avoided. Particle sizes of less than 2 μm are to be preferred. Samples may be either ground by hand in an agate pestle and mortar or larger samples may be ground in a small vibratory ball mill. However, it is difficult to grind quartz in agate and contamination of samples by agate from the mortar is possible (Borka and Kristiansen[642]). The use of boron carbide is recommended as a suitable alternative. The samples are best ground under a few drops of absolute alcohol, as it has been found that many minerals, particularly those containing OH groups in the lattice, can suffer alteration by vigorous dry grinding. The powdered material must be presented to the spectrometer in a diluted form, and many ways of achieving this have been suggested. Three methods have attained greater popularity than most and only these are considered in any detail.

PRESSED DISCS

A small sample (1–3 mg usually) is dispersed homogeneously in ca. 250 mg of a solid infrared transparent matrix and pressed into a disc. Cut-out frequencies of various infrared transparent solids are given in *Table 1.* Potassium bromide (or chloride) is most frequently used, although thallous bromide has been used to good effect with materials of higher refractive index (Smallwood and Hart[15]). Matching of the refractive indices of the

Table 1 CUT-OUT FREQUENCIES FOR VARIOUS INFRARED WINDOW MATERIALS (CUT-OUT POINT DEPENDS ON THICKNESS)

Window matrix	*Cut-out frequency* (cm^{-1})
KBr	ca. 310 (32 μ)
NaCl	ca. 500 (20 μ)
BaF_2	ca. 770 (13 μ)
CaF	ca. 1000 (10 μ)
CsI	ca. 170 (60 μ)
KRS 5	ca. 200 (50 μ)

sample and matrix is important if energy scattering effects are to be avoided. The sample and matrix are intimately ground together, again either by hand or in a ball mill. It is important to ensure that disc samples are not contaminated by mulling agents from previous preparations and it is a good idea to keep separate pestles and mortars for discs and mulls. Fusion of the mixture in a warm die under vacuum and high pressure produces a solid disc, usually transparent. Over-concentrated discs can be readily diluted by breaking in half and regrinding with additional KBr.

When only a very small sample (a few mineral grains) is available, a full-sized pressed disc would yield too dilute a sample. Although scale expansion devices might help rectify this, a better remedy is to reduce the quantity of KBr required, by the use of a cardboard former. The disc is pressed as before, except that, now, the more concentrated KBr/mineral mixture is confined to the area of the disc exposed to the infrared beam, and wastage in the unused portions of the disc is avoided. Similar cardboard formers are available for the preparation of microdiscs to be used in conjunction with a beam condenser.

Discs have the advantage over mulls in that the concentration and thickness of the specimen are readily determined, thus permitting their use in quantitative analysis. Should the mineral be required for further studies, and be insoluble in water, it can fairly readily be recovered from the KBr disc by dissolution of the latter.

It is important to bear in mind that ion-exchange reactions between the sample and the alkali–halide may occur. Meloche and Kalbus[68] did not detect exchanges in carbonates, phosphates, halides and some sulphates, but noted that exchanges appeared to have taken place in some nitrates and sulphates. Polyethylene is used as the support matrix in the preparation of pressed discs for the far-infrared examination of inorganic materials (Karr and Kovach[11]).

MULLS

The sample is dispersed in a liquid infrared transparent medium. Liquid paraffin (Nujol) is the most commonly used mulling agent, since the number of interfering absorption bands (due to the paraffin) are small. When the C—H stretching region (obscured by the liquid paraffin absorption) is important to the mineralogist either hexachlorobutadiene or fluorolube (a mixture of fluorinated hydrocarbons), which are both absorption free in that region, may be used. Interfering bands are given in *Table 2*. A larger sample than that required for the pelleting technique is taken and the comments regarding sample purity are especially pertinent here.

Table 2 ABSORPTION BANDS OF COMMON MULLING AGENTS (INTENSITY DEPENDS ON THICKNESS)

Mulling agent	*Absorption bands*
Nujol	3400–2750 cm^{-1}
	1800–1350 cm^{-1}
	with a small window between 1430 and 1380 cm^{-1}
	ca. 725 cm^{-1}
Fluorolube	1725–1585 cm^{-1}
	unusable below 1375 cm^{-1}
Hexachlorobutadiene	1640–1530 cm^{-1}
	1190–1150 cm^{-1}
	unusable below 1000 cm^{-1}

The mull is prepared by grinding 5–10 mg of sample with 1 drop of the mulling agent in either an agate pestle and mortar or a ball mill. After transference to an alkali–halide plate the mull is covered with a second plate, which forces the mixture to spread as a thin film. Ion-exchange effects with the cell plates are not unknown and must be considered when interpreting the spectrum.

SOLID DEPOSITS

(*a*) Finely ground minerals are deposited from either absolute alcohol or isopropyl alcohol on to infrared transparent plates (see Hunt *et al.*[33]), but the method suffers from excessive energy scattering in the wavelength regions below 2500 cm^{-1}. This method of preparation can lead to preferred orientations in the sample and hence atypical spectra, which is to be avoided in mineral identification. However, when preferred orientations are required, e.g. when using an infrared polariser for structure studies (Bilton, Gilson and Webster[645] and Grim and Kulbicki[672]), then this preparative technique is often essential. The advantage of this technique is that ion-exchange effects are minimised and interfering absorptions are avoided.

(*b*) Low melting solids may be melted and pressed into a thin film between two plates, but the application of this technique is naturally restricted.

(*c*) Szymanski and Peller[29] have described a sublimation technique for obtaining some powder deposits.

OTHER TECHNIQUES

Reflectance techniques have been employed since the late 1930s by Matossi *et al.*[17,18], but more recently by Ulrich[16], Lyon and Burns[28] and Vierne and Brunel[530,533]. The possible use of spectral emission for the investigation of minerals has been considered by Lyon[1] and Low[361]. Serratosa *et al.*[19,20] have described the use of a paper of orientated dried clay to obtain spectral absorption data from the O—H vibrations of clay particles. Infrared spectra of aqueous solutions have been obtained; the technique has been reviewed by Goulden[171]. References to works with single crystals have been made as and when they occur in Part III, but a theoretical treatment is given by Raman[30]. Ahlijah and Mooney[418] have employed the A.T.R. technique to phosphate solutions.

QUANTITATIVE ANALYSIS

Many authors have shown that in spite of the spectra having mainly broad overlapping absorption bands, quantitative analysis of selected mineral systems is possible. These analyses vary from the determination of single minerals, e.g. chrysotile in airborne dusts (Gadsden *et al.*[22]) to analysis of more complex mixtures.

A detailed account of the theory of quantitative analysis can be found in many of the standard infrared texts, e.g. Stewart[184], but a brief discussion of the principles and the hazards of quantitative analysis as applied to solid samples is included here.

Quantitative analysis by means of infrared absorption spectroscopy is based on the use of Beer's law:

$$A = \log \frac{1}{T} = abc$$

where A is absorbance, T transmittance, a absorptivity, b sample thickness, and c sample concentration.

$$\text{Transmittance} = \frac{\text{Transmitted radiation intensity}}{\text{Incident radiation intensity}} = \frac{I}{I_0}$$

Due to scattering by the solid samples and where partially overlapping bands are encountered, it is sometimes difficult to establish I_0. In these cases the baseline method of determining I_0 is employed; a straight line being constructed across the base of the band, to which baseline, at the frequency of peak maximum, the background (incident) radiation I_0 is measured.

Scattering of radiation by solid samples can lead to errors in intensity measurements. These errors could be eliminated by dissolution of the sample and the use of fixed path-length liquid cells. But if this should not prove possible, as is frequently the case, then discs or mulls must be made from finely ground solid. The major difficulty encountered with the use of discs and mulls in infrared quantitative analysis is in ensuring homogeneity within the matrix. Some compensation for inhomogeneity can be achieved by rotating the disc through 60° and averaging the three results obtained.

Accuracy may further be improved by the addition of an internal standard to the mixture before pressing the disc, and plotting absorbance (sample)/absorbance (standard) versus concentration, rather than plotting the absorbance of the sample versus concentration. The internal standard is chosen to provide a useful absorption band in a convenient spectral region, but with no interference with the absorption band in the sample.

Typical examples of the use of infrared spectroscopy in quantitative analysis in mineral and inorganic systems, serving as useful data to the methods and techniques employed, are given in references 674–685. The titles shown in parentheses are the present author's.

PART II

Spectra-Structure and Group Frequency Correlations

A group of atoms (usually three or more) possess $3N-6$ (or $3N-5$ for linear molecules and ions) normal modes of vibration, where N is the number of atoms in the group. Not all of the normal modes of vibration are infrared active (i.e. absorb infrared energy), since for them to be so, a change in the dipole moment must occur during the vibration. Those vibrations not present in the infrared spectrum because of a zero dipole moment change are said to be 'forbidden' and are found in the Raman spectrum. Hence the current interest in the re-emergence of the Raman technique, brought about by the advent of the laser (see, for example, references 509, 518, 600).

When considering the number of absorption bands to be found in the spectrum, the symmetry of the molecule or ion must be considered. For highly symmetrical molecules or ions the number of absorption bands in the spectrum is less than at first expected from the substitution of N in the formula $3N-6$ (or 5). In these cases it can be seen that two or more of the expected vibrations are identical with one another and absorb at the same frequency. When two vibrational modes are identical the mode is said to be doubly degenerate and when three identical modes exist, triply degenerate. Specific groups of atoms are dealt with in more detail later in the text. Added to this is the complication that some vibration frequencies may be so close as to be unresolved, further decreasing the number of bands observed. Lowering of the symmetry of the group, as in the case of distortion of the structure and/or upon co-ordination to the cation, removes the degeneracy. This results in the splitting of the degenerate absorption into two or three bands and often the appearance of weak 'forbidden' absorption bands. The distinction between a split fundamental and a pair of unresolved absorption bands may not be easy, but a clue is usually afforded by the presence of the 'forbidden' absorption band and the splitting of other degenerate fundamentals.

In addition, the spectra from crystalline (as opposed to truly amorphous) compounds exhibit absorption bands arising from lattice modes—vibrations due to translatory and rotatory motions of a molecule or ion in the crystalline lattice. These absorptions usually occur at frequencies below 300 cm^{-1} but, because of possible combination with the fundamental modes, they may give rise to absorption bands at higher frequencies. Thus the vibrational spectrum obtained from solid inorganic materials contains considerable information, all of which may not be instantly recognisable. However, if the data are interpreted with caution, not only is identification of an unknown possible, but also information concerning the structure, normally only obtained after rigorous X-ray structure analysis, is available.

Should a molecule or anion have several probable structures, each of which belongs to a different point group (i.e. differing symmetry), then the number of infrared (and Raman) active fundamentals should be different for each structure. For a more detailed discussion on symmetry and molecular vibrations see the relevant chapters in Hertzberg[26] or Wilson, Decius and Cross[25]. Thus, the most likely candidate will be the one which predicts the observed number of absorption bands, but caution must be exercised because of the problems mentioned above. Examples of studies into the effects of symmetry modifications upon the appearance of the infrared spectrum are quoted in the discussions in the following paragraphs, where relevant. Particularly noteworthy are the cases of the carbonate and sulphate minerals.

Review articles on minerals and related inorganic infrared spectra include those by Setkina[38], Achmanova[5], Tai[274], Ferraro[488,489] and Kalbus[315], whilst Karr[477] alone and with Kovach[11] and Feairheller and Miller[610] have paid particular attention to the far-infrared region.

Since the absorption frequency is dependent upon the nature of the atoms involved in the vibration, a range of absorption frequencies exists for different species. However, for a given species, a few absorption bands characteristic of the group can be recognised. The frequency of a given vibration is constant to within reasonably definable limits but varies within those limits due to the environment within which the group finds itself. As already mentioned, structural modification due to co-ordination with cations of differing mass and charge affect the vibration frequency, but it is the aforementioned constancy that has led to the concept of 'group frequencies' and their application to identification problems.

By reference to correlation charts it is usually possible to identify the constituent groups of an inorganic compound, although direct comparison with well-characterised reference standards is essential before a claim to positive identification is made. Frequently, because of the non-stoichiometry of so many mineral species, no exact reference spectrum will be available, but the unknown can still be placed in its correct place within a mineral group. A correlation chart for polyatomic inorganic ions is displayed at the start of Part III (p. 36). (*Figure 6*). A careful scrutiny of the minerals, etc., contained in the relevant section of Part III could either identify the unknown or at least give a clue to its identity.

The factors affecting the characteristic group frequencies are summarised below, according to the number of atoms in the anion, with the exception of the first section which deals with binary compounds only. A review by Karr[477] covers the recent work in the far-infrared region and some of the findings quoted are mentioned briefly here, but many anionic species are not represented and further work in this region is still required.

BINARY COMPOUNDS

HALIDES

Diatomic halides are well known for their transparency through large regions of the infrared spectrum, hence their adoption as matrices in the pelleting technique and as infrared transparent windows for mull and solution cells. The halides do, however, have strong infrared absorptions, although the number and location depends on the elements involved. A good selection of halide spectra is included by Nyquist and Kagel[4]. Godnev *et al.*[82] have calculated the normal vibration frequencies of several zirconium halides. Rare earth fluorides have been shown by Batsanova *et al.*[248,410] to exhibit three infrared absorptions in the regions 500–400 cm^{-1}, 400–360 cm^{-1} and 290–270 cm^{-1}. The spectra of mercury, bismuth and antimony halides have been recorded by Glasner and Reisfeld[185] but they warn against the use of the pressed disc technique for mercuric chloride and bromide, antimonous chloride and bismuth chloride. Both absorption and reflection spectra of ammonium halides have been recorded by Bovey[254]. A useful and extensive discussion of the far-infrared spectra of the halides is given by Karr[477].

OXIDES

McDevitt and Davidson[382] have obtained the infrared lattice vibration spectra in the region 700–50 cm^{-1}, of eight cubic rare earth oxides. From the data they obtained it was evident that the frequency position was determined by the crystal structure and the results clearly showed the effect of the lanthanide contraction in the series. Oxides are usually non-absorbing above 1000 cm^{-1} (silica is not a simple oxide) but most show a strong band and some weaker ones in the 900–300 cm^{-1} region. Other lattice vibration modes have been found below this region. Very full lists of oxide absorption maxima may be found in McDevitt and Baun[416] and a large number of spectra in Nyquist and Kagel[4]. The infrared transmission curves of several oxides are given in the region 4000–1000 cm^{-1} by Wickersheim and Lefever[450]. Karr[477] has given a very full account of the far-infrared spectra of a few simple oxide minerals.

SULPHIDES, SELENIDES, TELLURIDES AND ARSENIDES

Selenides and tellurides have been grouped together with sulphides, since studies usually include all three. Any sulphide absorption bands that do arise usually occur below 400 cm^{-1} but there is no simple correlation. A single intense absorption is quoted by Karr and Kovach[11] for calcium, strontium and barium sulphides in the 255–175 cm^{-1} region, the frequency of the vibration decreasing with increasing atomic weight of the cation. Mitsuishi *et al.*[200] have found only one infrared absorption due to the lattice vibrations of zinc and cadmium sulphides, selenides and tellurides—that occurring between 285 and 140 cm^{-1}. Absorption bands for selenides and tellurides are to be found below 400 cm^{-1}; several examples of both are available in Nyquist and Kagel[4] who also include 17 sulphides. Sulphide minerals have been studied by Gillieson[545], and sulphide, arsenide and telluride minerals by Povarennykh *et al.*[576].

DIATOMIC IONS

For an *N*-atomic linear compound or anion there are $3N-5$ fundamental vibrations. Diatomic anions have $N = 2$ and there is therefore only one fundamental vibration, that of stretching. Whereas in homopolar anions (X–X) no dipole moment change is effected, in hetero-polar anions (X–Y) this vibration is infrared active and gives rise to a single fundamental absorption peak. Other peaks, particularly in the far-infrared region, may be present due to lattice vibrations, but it is the stretching vibration which is quoted as the 'group frequency'.

CYANIDES $(CN)^-$

A sharp strong absorption band due to the C≡N stretch in the region 2250–2050 cm^{-1} is characteristic of the cyanide ion. Nyquist and Kagel[4] have included six cyanides in their collection and Ahlijah and Mooney[418] have used the attenuated total reflectance technique to obtain information from cyanide solutions.

HYDROXIDES $(OH)^-$ (and water)

The O—H stretching vibration gives rise to a strong sharp band in the region 3700–3500 cm^{-1}. However, the stretching vibration in hydroxides is partnered by a bending vibration.

Since an M—O band of one type or another is formed, the oxygen atom is fixed relative to the cation, leaving the hydrogen free to vibrate about the oxygen in a bending mode. The 'free' hydroxyl stretching vibration is distinguishable from the O—H stretching vibration in water by the reduced frequency and increased broadening of the latter, due to hydrogen bonding. Another distinction between 'free' hydroxyl and both absorbed water and water of hydration, is that the O—H bending frequency occurs in the 1200–600 cm^{-1} region in the former but near 1650 cm^{-1} in the latter cases.

Duval and Lecomte[348] have given a detailed interpretation of hydroxides of the type XOH, $X(OH)_2$, $X(OH)_3$ and $X(OH)_4$. The relationship between the O—H distance and the O—H stretching and M—O—H bending frequencies has been discussed by Glemser and Hartert[182,197] (who have also reported the spectra of a number of hydroxides[243]). This relationship has been reinvestigated by von Schwarzmann[471] and Wickersheim[154]. Buchanan[147] has investigated the near-infrared spectra of crystalline alkali hydroxides.

A distinctive strong-to-medium, broad, multi-peaked absorption band over the range 3400–2000 cm^{-1}, assignable to the OH stretching frequency, is characteristic of the very acidic protons of acids and the corresponding salts of the type HSO_4^-, HPO_4^{2-}.

It seems pertinent at this point to consider the absorption bands that might arise due to water present either as 'absorbed water' or 'water of hydration' (crystallisation). Bent triatomic molecules (see page 18) have three normal modes, the symmetric and asymmetric stretches and the bending vibration. All three vibrations are infrared active, but the symmetric and asymmetric stretching modes are not usually resolved in 'free' or 'absorbed' water, due to the effects of hydrogen bonding. Thus the two stretching vibrations are apparent as a single broad absorption band in the 3400–3200 cm^{-1} region. The bending vibration gives rise to a weaker absorption in the 1650–1600 cm^{-1} region. However, in the case of 'water of hydration' there is less chance of hydrogen bonding effects, and usually the two stretching vibrations are observed (although not normally well resolved) in the 3600–3200 cm^{-1} region, and the H—O—H bending vibration generally occurs near 1650 cm^{-1}.

Many studies into the nature of the bonding in hydrates have been made. The absorption spectra of several hydrates have been examined by Glemser and Hartert[244], Lucchesi and Glasson[320] and Bellamy *et al.*[557]. Lecomte and Duval[290] and Lyon and Kinsey[289] have been particularly interested in the problem in relation to mineral species. Detailed investigations into the nature of the bonding in hydrated metal oxides have been carried out by Glemser[86] and Cabannes-Ott[321].

Comments on the far-infrared spectra of hydrous oxides have been made by Karr[477].

TRIATOMIC IONS

Linear YXY ions

URANYL $(UO_2)^{2+}$

Very few ions of the YXY type are important to the mineralogist but perhaps the uranyl ion $(UO_2)^{2+}$ is worthy of a mention. For linear ions there are $3N-5$ fundamental vibrations; in the case of linear triatomics four vibrations are expected.

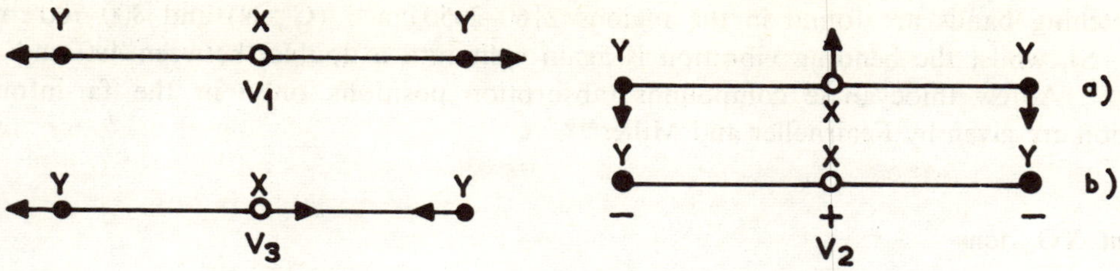

Figure 1. Normal modes of vibration in linear triatomic molecules/ions

As can be seen in *Figure 1*, ν_2, the bending modes, *a* and *b* are degenerate and differ only in the direction of the bend; the vibration frequencies are identical. The symmetric stretch ν_1 should be infrared inactive; however, a weak absorption is found indicating that the $(UO_2)^{2+}$ ion is slightly bent. Jones and Penneman[478] quote the following absorption bands for the uranyl ion:

	ν_1	ν_2	ν_3
$(UO_2)^{2+}$	880–860 cm^{-1}	210–199 cm^{-1}	960–930 cm^{-1}

Some spectra and structures of simple $(UO_2)^{2+}$ salts are given by Gatehouse and Comyns[242] but these results are disagreed with by Allpress and Hambly[114]. The U—O bond distance in uranium compounds has been determined by Jones[343,491] and Ohwada[414]. Prigent[223] has investigated the anhydrous uranyl halides and confirmed that the anhydrous uranyl ion is linear. Several uranium salts have been studied by Lecomte and Freymann[189].

Linear XYZ ions

As with linear YXY molecules and anions, the XYZ group also have three fundamental vibration modes, the ν_2 bending mode again being doubly degenerate, but ν_1 is now infrared active (see *Figure 1*). However, the crystal field removes the degeneracy in cyanates and thiocyanates and two ν_2 bands are found.

CYANATES $(NCO)^-$

Absorption bands corresponding to the symmetric and asymmetric stretching frequencies are observed at 2250–2160 cm^{-1} and 1300–1200 cm^{-1}. These absorption frequencies approximate to the C≡N and C—O stretching vibrations, respectively. The normally doubly degenerate bending vibration is split and occurs in the 650–590 cm^{-1} region. A comparison of the vibration frequencies of the cyanate ion and some isoelectronic molecules and ions has been made by Waddington[267].

THIOCYANATES $(SCN)^-$

Jones[94] and Pecile *et al.*[427] have investigated the spectra of various thiocyanates. Thiocyanato and selenocyanato complexes have been studied by Pecile[389]. The ν_1 and ν_2

stretching bands are found in the regions 2160–2050 cm^{-1} (C≡N) and 800–700 cm^{-1} (C—S), whilst the bending vibration is again split into a doublet between 490 and 400 cm^{-1}. A few thiocyanate compounds (absorption positions only) in the far-infrared region are given by Feairheller and Miller[610].

Bent XO_2 ions

Non-linear polyatomic molecules and ions have $3N-6$ normal modes of vibration. For bent triatomic groups there are three fundamental modes, all of which are infrared active whether the group is symmetrical or not. There are two stretching vibrations, the symmetric and the asymmetric, but only one bending vibration, all of which are shown in *Figure 2*.

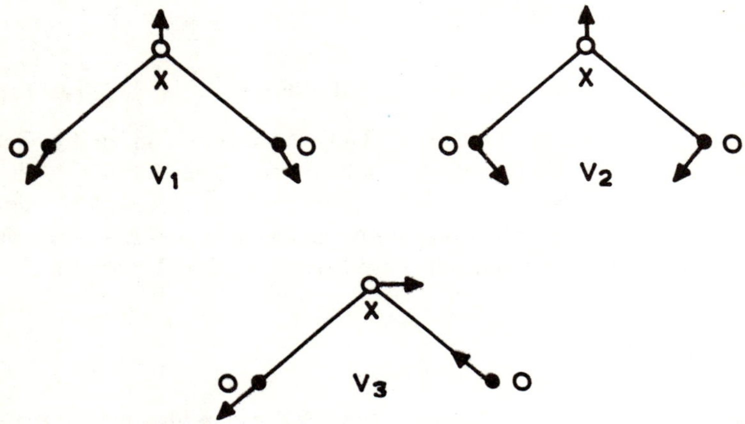

Figure 2. Normal modes of vibration in bent triatomic molecules/ions

A study of ions of the type $(XO_2)^-$ has been made by Duval and Lecomte[163,176] and Duval, Lecomte and Morandat[237], with particular reference to metaborates, chlorites and nitrites.

METABORATES $(BO_2)^-$

Studies by Duval and Lecomte[163] have led them to believe that polymerisation of the metaborate ion $(BO_2)^-$ occurs with time. If this is the case, then the available absorption data for metaborates must be open to some doubt unless obtained from fresh material. References to work with metaborates have been included in the section dealing with orthoborates.

CHLORITES $(ClO_2)^-$

Absorption bands due to the asymmetric and symmetric stretching vibrations are to be found at 860–840 cm^{-1} and 790–700 cm^{-1}, respectively, with the bending vibration at 450–400 cm^{-1}. Some absorption frequencies for a few chlorites are given by Duval and Lecomte[237].

NITRITES $(NO_2)^-$

The two stretching vibrations occur in the 1365–1320 cm^{-1} and 1270–1210 cm^{-1} regions, but the absorptions are frequently so broad as to be overlapping. Duval *et al.*[237] have examined four nitrites, and four nitrites have also been included by Nyquist and Kagel[4] in their collection.

FOUR ATOM IONS

Pyramidal XO_3 ions

Four atom non-linear ions or molecules have six possible normal modes of vibration, but two of these vibrations are doubly degenerate and hence only four fundamental vibrations are found. These four vibrations, both infrared and Raman active, are depicted in *Figure 3*.

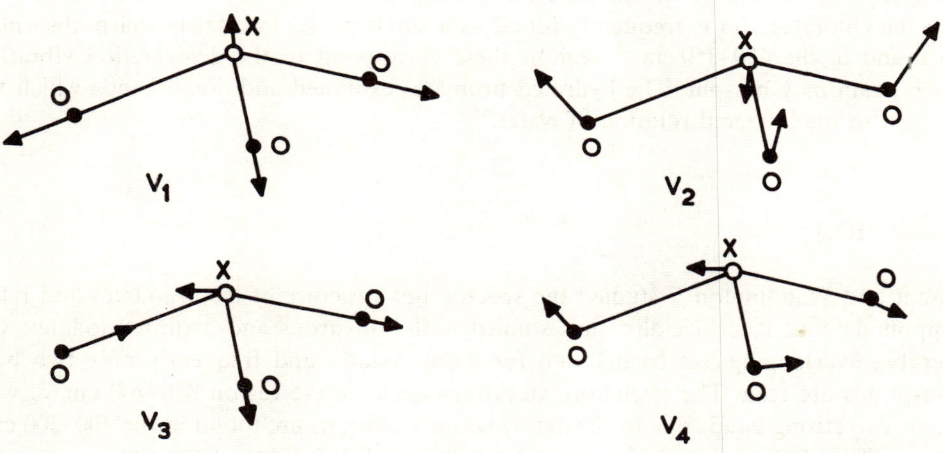

Figure 3. Normal modes of vibration of pyramidal XO_3 molecules/ions

Although four absorption bands are expected, the two non-degenerate symmetric vibrations, ν_1 and ν_2, occur close to each other and are usually observed as one strong broad band, often exhibiting a shoulder. However, the symmetry of the XO_3 ion may be lowered upon co-ordination to a cation and the degeneracy of the ν_3 and ν_4 vibrations will be removed. The resulting split 'degenerate' vibrations may or may not be resolved, depending upon the magnitude of the frequency difference. A discussion of the spectra of inorganic compounds containing the XO_3 group is given by Rocchiccioli[302], including chlorates, bromates, iodates, sulphites, bisulphites, pyrosulphites, selenites, hydroselenites and pyroselenites.

The group of anions, chlorates, bromates and iodates, naturally, exhibit similar spectral features, but the increasing mass of the halide atom brings about an overall reduction in frequency of the vibrations as the group is descended.

CHLORATES $(ClO_3)^-$

A number of metal chlorates have been examined by Rocchiccioli[214] and four are included in Nyquist and Kagel[4]. The stretching vibrations ν_1 and ν_3 occur together in the region 1050–900 cm^{-1} as a broad band, often exhibiting either two peaks or an unresolved shoulder, and the deformation vibrations ν_2 and ν_4 occur at 630–600 cm^{-1} and 525–475 cm^{-1}, respectively, and may be split as a result of the removal of the degeneracy by a lowering of the symmetry.

BROMATES $(BrO_3)^-$

Rocchiccioli[350] has obtained the infrared spectra of a number of anhydrous and hydrated metallic bromates. A broad absorption between 880 and 740 cm^{-1} has been attributed to the stretching vibrations, but the absorption band in bromates tends to be less broad than with the chlorates and is frequently found as a single peak. Two fairly sharp absorptions are found in the 450–350 cm^{-1} region; these correspond to the deformation vibrations, which again may be split. The hydrated bromates exhibited additional bands which were assigned to the hindered rotation of water.

IODATES $(IO_3)^-$

Dasent and Waddington[132] studied the spectra and structure of metal iodates and related compounds, and Rocchiccioli[349] has studied both anhydrous and hydrated iodates. Considerable overlapping has been found for many iodates and frequently only two broad absorptions are seen. The stretching vibrations occur in the region 830–700 cm^{-1}, whilst one or two strong bands due to the deformation vibrations are found in the 400–300 cm^{-1} region. These latter vibrations may be split due to the removal of the degeneracy and it has been suggested that extra bands observed at 480–420 cm^{-1} may be due to the metal–oxygen vibration. Hydrated iodates, like the corresponding bromates, show a supplementary band which has been assigned to the hindered rotation of water.

SULPHITES $(SO_3)^{2-}$ AND SELENITES $(SeO_3)^{2-}$

A single broad absorption, in the region 1090–920 cm^{-1} is observed, for the ν_1 and ν_3 stretching frequencies of sulphates are unresolved. (It is interesting to note that in this case the symmetric stretching frequency is higher than the asymmetric stretching frequency

—see Adler and Kerr[257] for XO_4 examples.) Both the ν_2 and ν_4 degenerate deformation bands are frequently split and for sulphites are to be found in the 660–600 cm^{-1} and 520–450 cm^{-1} regions, respectively. As can be seen, the sulphite and chlorate absorption bands occur within the same spectral regions and can be confused. However, chlorates do absorb at slightly lower frequencies than the corresponding sulphites, but differentiation between these two ions is difficult unless the cation is known and direct comparison with reference spectra can be made. Rocchiccioli[69,135], Newman and Powell[124] and Nyberg and Larsson[648] have examined the infrared spectra of a number of metal sulphites.

A sharp strong shoulder is usually found on the high frequency side of the selenite stretching absorption band which occurs in the 850–700 cm^{-1} region. Splitting of the ν_2 and ν_4 bands is not always apparent, but the doublet due to these vibrations is found in the region 550–350 cm^{-1}. Caillere and Pobeguin[63], Rocchiccioli[222,302], Sathianandan *et al.*[309] and Petrov *et al.*[583] have investigated the infrared spectra of a number of anhydrous and hydrated selenites.

ARSENITES $(AsO_3)^{2-}$

Paillette, Lankoande, Hadni *et al.*[137] have studied the infrared spectra of alkaline earth arsenites and found absorption in the range 840–700 cm^{-1} common to all those compounds investigated.

TITANATES $(TiO_3)^{2-}$, STANNATES $(SnO_3)^{2-}$, NIOBATES $(NbO_3)^{2-}$ AND TANTALATES $(TaO_3)^{2-}$

Last[231] has based his interpretations of the vibrational spectra of metal titanates on the distorted TiO_6 unit, and has shown that the perovskite niobates and titanates exhibited similar spectra. Hadni and Michel[238] have shown that the spectrum of barium titanate obtained from a crystal is markedly different to that obtained from a powdered sample. Nyquist and Kagel[4] have included several titanates, niobates, stannates and zirconates in their collection of infrared spectra. Inorganic compounds within this group generally exhibit broad absorption bands in the region 800–150 cm^{-1}, often with a 'window' between 500 and 400 cm^{-1}. Tarte[128] has studied the orthotitanates $(TiO_4)^{4-}$, and stannates have been included by Duval and Lecomte[67]. Niobates, tantalates and antimonates of various structures have been studied by Rocchiccioli *et al.*[541,565,566,581].

Planar XO_3 Ions

There are six normal modes of vibration for the planar XO_3 ion: the symmetric stretch ν_1, which is infrared inactive, the out of plane bend ν_2, the asymmetric stretch ν_3 and ν_4, the planar bend, the latter two being doubly degenerate. These four modes are depicted in *Figure 4*.

Since planar XO_3 ions have only three infrared active vibrations they are theoretically distinguishable from their pyramidal counterparts. However, as the stretching vibrations

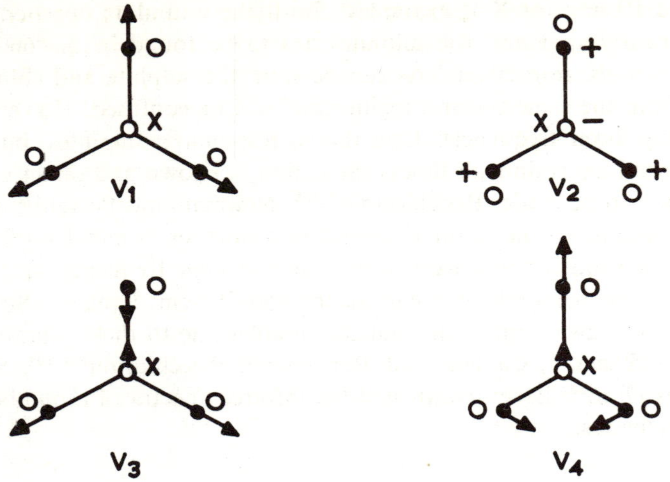

Figure 4. Normal modes of vibration of planar XO_3 molecules/ions

of the pyramidal XO_3 ions are frequently observed as a single broad absorption band, the distinction is thus made much more difficult.

Crystal structure plays a very important part in deciding the appearance of the spectra of the species within this group.

For example, calcium carbonate exists in three structural forms, named after the corresponding minerals, calcite, aragonite and vaterite, all having the same chemical composition. A different spectrum is observed for each of the structural forms. Numerous discussions on the relationship between spectra and structure in this series have taken place in the literature, not least among these being the contributions, especially of carbonate structures, by Ross and Goldsmith[371,377] and by Adler and Kerr[65,313]. Aragonite, calcite and vaterite structures have been studied, by means of their infrared spectra, by Weir and Lippincott[59].

In calcite-type structures the XO_3 ion should have D_3 symmetry. The site symmetry in vaterite is unknown, but in aragonite-type structures the symmetry is C_s. As there is no change in the selection rules governing D_{3h} and D_3, then XO_3 ions having the calcite structure should have the spectrum of the unperturbed ion. However, in aragonite structures the symmetric stretch ν_1 will become infrared active and the vibrations ν_3 and ν_4 will be split. These predictions as to the appearance of the calcite and aragonite spectra are borne out in practice. It has been found that the overlap or coincidence of bands, however, often limits the usefulness of the spectra for the purposes of elucidating molecular symmetry in compounds with ions of this type. Compounds with the vaterite structure give spectra diagnostically different from either of the other two types, but have not yet been structurally determined and further work is needed.

BORATES $(BO_3)^{3-}$

Infrared absorption spectra of both meta and orthoborates (BO_2^{2-} and BO_3^{3-}) have been studied in minerals by Moenke[73,120] and as metal salts by Duval and Lecomte[67].

Steele and Decius[93] have made assignments for the orthoborate anion from the spectra they obtained from the borates of lanthanum, scandium and indium. The infrared spectra of borates, either of the rare earth elements or of trivalent cations, with the calcite, aragonite or vaterite structures, have been investigated by Laperches and Tarte[417], and the rare earth borates have been studied by Denning and Ross[575]. Boron (and beryllium) minerals have been recorded by Povarennykh and Nefedov[571]. Because the borate ion absorption frequency depends on both its environment and structure, it is not possible to state the expected 'group frequency' range with absolute certainty.

However, recognition of the presence of borate is possible. Absorptions in the regions 1350–1250 cm^{-1} (very strong and broad) in the case of the 'calcite'- and 'aragonite'-type structures, and a complex broad absorption between 1100 cm^{-1} and 900 cm^{-1} in the case of 'vaterite'-type structures, may be regarded as good evidence for borates. The ν_2 and ν_3 absorptions occur in the 800–600 cm^{-1} region both for calcite and aragonite types, and ν_3 may be split. In aragonite types the ν_1 symmetric stretch becomes infrared active and occurs between 900 and 850 cm^{-1}, whilst vaterite-type structures are recognisable by the dearth of absorption bands in these regions. Naturally, the spectra of borate compounds are not unlike those from the corresponding carbonates, but the absorptions generally occur at lower frequencies. In many minerals the borate ions are not isolated, but occur as condensed polyatomic ions. Space prevents discussions on this aspect of the structure and spectra of borates and the reader is referred to the work of Takeuchi[145] who has discussed at some length the structures of the borate polyatomic ions found in many minerals.

CARBONATES $(CO_3)^{2-}$

Carbonates show strong infrared absorptions in the regions 1530–1320 cm^{-1} (ν_3), usually restricted to 1450–1400 cm^{-1}, medium strength bands between 890 and 800 cm^{-1} (ν_2) and at 760–670 cm^{-1} (ν_4). When present, as in aragonite-type carbonates, the symmetric stretch ν_1 occurs in the 1120–1040 cm^{-1} region. As already stated, the ν_3 and ν_4 modes are doubly degenerate and under certain symmetry conditions will split into doublets (although usually this is only observed for ν_4) and the ν_1 stretching frequency becomes apparent. The effects of symmetry and structure on the spectra of carbonate minerals has been discussed in some detail by Adler and Kerr[313]. The results have enabled them to recognise, or at least suspect, the presence of non-identical carbonate ions in the unit cell, a task normally only accomplished by rigorous X-ray structure analysis. Many low frequency absorption bands have been observed and these have been discussed by Karr[477], who has reported that the frequencies of certain lattice modes decrease with increasing mass and radius of the cation for both the calcite- and aragonite-type structures, and that the major and minor far-infrared absorption bands appear to be diagnostic for each carbonate mineral.

In bicarbonates the symmetry is lowered to C_{2v} and all six expected normal modes for the carbonate group become active, plus the stretching mode of the hydroxyl group. Absorptions are found at 1700–1600 cm^{-1} (strong broad), near 1400 cm^{-1} (very strong), between 1000 and 950 cm^{-1} (medium, two or three bands), in the 850–800 cm^{-1} region (medium sharp) and between 720 and 650 cm^{-1} (medium, one or two bands).

Moenke[494], Caillere and Pobeguin[61,63], Louisfert[239] and Duval, Duval and Lecomte[37] have examined metallic carbonates, whilst the carbonates of the rare earth elements have been studied by Akhmanova and Orlova[399] and Caro *et al.*[449]. Apart from the work of Adler and Kerr[313], already mentioned, the spectra of carbonate minerals have been obtained by Huang and Kerr[110] and Elderfield and Chester[549], and in the far-infrared region by Angino[400]. Carbonate and nitrate minerals have been studied by Boldyrev and Povarennykh[531]. Carbonates in the visible and near infrared have been studied by Hunt and Salisbury[542].

NITRATES $(NO_3)^-$

The symmetry of the free nitrate ion is D_{3h}, but in ionic metal nitrates the symmetry is progressively lowered as the metal changes from monovalent to tetravalent. These findings by Ferraro[474] are similar to the observations of Addison and Gatehouse[97]. An absorption band is found in the region 1500–1300 cm^{-1} for the asymmetric stretch ν_3 and the two bending vibrations ν_2 and ν_4 occur between 850 and 800 cm^{-1}, and 770–715 cm^{-1}, respectively. Trivalent metal nitrates included by Nyquist and Kagel[4] show distinct splitting of the ν_3 stretching frequency to give two strong broad bands between 1550 and 1250 cm^{-1}, the emergence of the symmetric stretch ν_1 at 1070–1010 cm^{-1} and also the splitting of the ν_4 band. The infrared spectra of metal nitrates have been recorded by Vratny[72] and in the far-infrared region by Ferraro and Walker[353] and Feairheller and Miller[610]. Two rotational and two translational bands are found for the alkali–metal nitrates in the far-infrared region. Karr[477] has shown that, as with carbonates, the increased mass of the cation decreases the frequency of the two translational modes. Spectra from single crystals of metal nitrates have been obtained by Hafele[283]. Distinction between nitrate and nitrite is often afforded by the presence of weak sharp bands in the 1800–1700 cm^{-1} region in the former.

FIVE ATOM IONS

Square planar anions and inorganic co-ordination chemistry are inseparable and since the latter is outside the scope of this book the reader is referred to Nakamoto[8] for a detailed account of the group frequencies of square planar molecules, anions and the spectra of inorganic co-ordination compounds generally. To the mineralogist the most important group of anions are the tetrahedral five atom anions of the type XO_4, where X = P, S, Si, As, Mn, W, V, Cl, Cr, Se, Mo, etc.

Nine normal modes of vibration ($3N-6$) are expected for the five atom anions. However, as we have already seen, if during a symmetry operation one vibration of a particular mode is indistinguishable from a second, then the vibrational frequencies will be identical and the mode is referred to as being doubly degenerate. Similarly, we may have triply degenerate vibrations when three modes are equivalent but this is only possible if the molecule or ion possesses a threefold axis of symmetry. This threefold axis of symmetry is restricted to the tetrahedral and octahedral structures, both of which do exhibit triply degenerate vibrations. Thus we may have non-degenerate, through doubly to triply degenerate vibrations present and the tetrahedral ion has all three types of vibration present. *Figure 5* illustrates the four normal modes of vibration of the tetrahedral XO_4 ion, the doubly degenerate ν_2, the non-degenerate stretching vibration ν_1 and two triply degenerate vibrations ν_3 and ν_4, the latter two being the only infrared active vibrations.

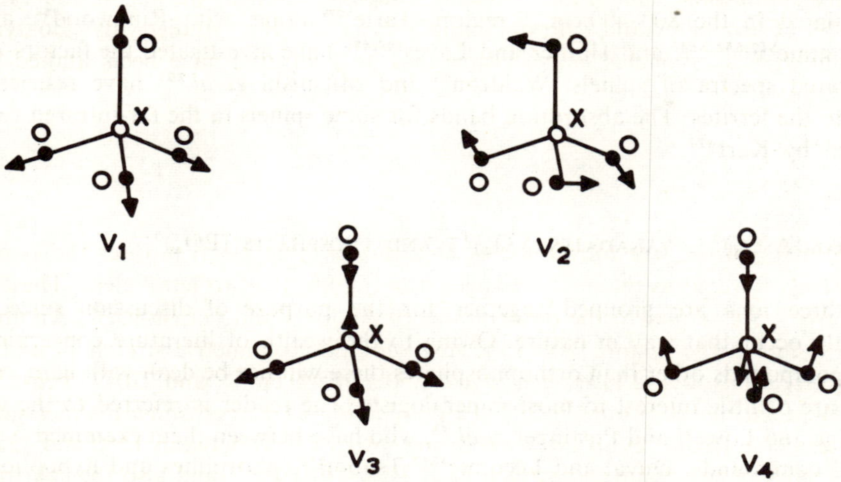

Figure 5. Normal modes of vibration of XO_4 molecules/ions

Distortion of the tetrahedron will lower the symmetry, causing the non-active vibrations ν_1 and ν_2 to absorb in the infrared, and removing the degeneracy of the infrared active ν_3 and ν_4 vibrations. Normally, the degenerate modes split observably into several components, two in the case of doubly degenerate and three in triply degenerate modes. In some cases, however, the splitting is not resolved in the spectra, either because of the low magnitude of the split, the broadness of the bands or the low intensity of one of the components.

This factor frequently accounts for the disparity in the number of bands (and the apparent frequency of their absorption maxima) reported by various investigators. The effects of distortion of the tetrahedron upon the spectra of the anions will be discussed, where they are important, for each of the anions in turn. Duval and Lecomte[181] and Guerchais *et al.*[447] have recorded the infrared spectra of several compounds containing the XO_4 group, and more recently Tarte and Thelen[561] have presented the spectra of $Ba_3(XO_4)_2$ and $Sr_3(XO_4)_2$, where X is P, As, V, Cr, Mn.

AMMONIUM $(NH_4)^+$

Before going on to deal with the anions, it is relevant at this point to consider the vibration frequencies of the ammonium tetrahedral cation. The asymmetric stretching vibration ν_3 occurs in the region 3350–3050 cm^{-1} and may be either obscured by or confused with the absorptions arising from the presence of water. However, ν_4 is found between 1450 and 1390 cm^{-1} and is a usefully diagnostic absorption. Other bands due to combinations and overtones are also observed. A study of the spectra, structure and the nature of hydrogen bonding in ammonium salts has been made by Waddington[241].

SPINELS

Spinels have the general formula A_2BO_4, with the A atoms occupying octahedral sites. Although the spinels are generally regarded as multiple oxides, they are included amongst the tetrahedral ions, since the B atoms occupy tetrahedral sites. Many spinels can be differentiated in the 800–400 cm^{-1} region. Tarte[122] alone, with Ringwood[44], and with Preudhomme[562,563,664], and Hafner and Laves[134,495] have investigated the factors affecting the infrared spectra of spinels. Waldron[55] and Mitsuishi *et al.*[207] have restricted their studies to the ferrites. The absorption bands for some spinels in the far-infrared region are discussed by Karr[477].

ARSENATES $(AsO_4)^{3-}$, VANADATES $(VO_4)^{3-}$ AND PHOSPHATES $(PO_4)^{3-}$

These three ions are grouped together for the purpose of discussion since they so frequently occur that way in nature. Owing to the wealth of literature concerning phosphorus compounds other than orthophosphates these will not be dealt with here, especially as they are of little interest to most mineralogists. The reader is referred to the works of Corbridge and Lowe[47] and Pustinger *et al.*[76], who have between them examined a very full range of compounds. Duval and Lecomte[131], Tsuboi[230] (phosphites and hypophosphites), Bergmann *et al.*[215] (pyrophosphates), Laforge-Kantzer[236] (poly and metaphosphates), Steger and Leukroth[284] (pyrophosphates), Mutschin and Maennchen[256,307] and Steger and Kabner[413] (diphosphates) have been more selective.

Hertzberg[26] places the ν_3 and ν_1 vibrations for the tetrahedral PO_4^{3-} ion at 1082 cm^{-1} and 980 cm^{-1}, respectively; however, the range is extended to 1120–940 cm^{-1} due to the effects of the cation and changes in crystal structure. Appearance of the ν_1 mode is indicative of a lowering of the molecular symmetry. Hezel and Ross[415] have discussed the factors affecting the infrared spectra of phosphates. The spectrum depends on whether the orthophosphate is combined with a mono, di, or trivalent metal. Chapman and Thirwell[308], Duval and Lecomte[150,152] and Steger and Schmidt[435] have discussed the spectra of orthophosphates. A detailed study of the effect of symmetry and substitution in the pyromorphite, mimetite and vanadinite series is given by Adler and Kerr[257,432].

There is a considerable similarity between the physical configurations of the AsO_4^{3-} and PO_4^{3-} absorption bands, whilst those of AsO_4^{3-} and VO_4^{3-} are quite different. Absorption ranges for the three groups are given in *Table 3.*

Table 3 ABSORPTION FREQUENCIES FOR THE ARSENATES, PHOSPHATES AND VANADATES

	ν_3	ν_1 ?	ν_4	
$(AsO_4)^{3-}$	850–770 cm^{-1}		550–400 cm^{-1}	
$(PO_4)^{3-}$	1120–940 cm^{-1}	970–920 cm^{-1}	600–400 cm^{-1}	often
$(VO_4)^{3-}$	900–700 cm^{-1}	ca. 870 cm^{-1}	410–300 cm^{-1}	split

It has been noticed that, frequently, as expected, the ν_3 frequency is greater than that of ν_1, but that occasionally the ν_1 frequency is the greater. For example, ReO_4^-, VO_4^{3-}, MoO_4^{2-}

and WO_4^{2-} have ν_1 on the high frequency side of the ν_3 absorption, but SO_4^{2-}, SeO_4^{2-}, PO_4^{3-}, SiO_4^{4-}, CrO_4^{2-}, MnO_4^-, ClO_4^- and IO_4^- have ν_1 on the low frequency side. Adler and Kerr[257] have noticed that the former tetrahedra all have large fourfold radii, whereas the latter have relatively small fourfold radii, but further investigation into this topic is required before positive deductions may be made. A few arsenates and vanadates have been included by Nyquist and Kagel[4]; the vanadates of Cr, Fe, Co, and Ni have been studied by Miller *et al.*[568] and the orthovanadates with the apatite structure by Baran and Aymonino[570]. A number of vanadate minerals have been recorded by Povarennykh[630].

SULPHATES $(SO_4)^{2-}$

The fundamental frequencies of the sulphate ion are given by Hertzberg[26] at 1104 cm^{-1} (ν_3), 981 cm^{-1} (ν_1), 618 cm^{-1} (ν_4) and at 451 cm^{-1} (ν_2). Absorption bands are usually found in the range 1210–1040 cm^{-1}. They are strong and may or may not be split; in the former case the range may be extended. Frequently, a doublet is found in the 680–570 cm^{-1} region, whilst a weak sharp absorption in the range 1030–960 cm^{-1} has been ascribed to coupling of molecular vibrations, by Nakamoto[475]. When the $(SO_4)^{2-}$ tetrahedron is distorted, ν_3 splits to give rise to two or three peaks, the resolution of which depends upon the magnitude of the split, and ν_1 becomes active usually near 1000 cm^{-1}. Similarly, the ν_4 vibration may be split into three absorption bands. Hezel and Ross[415] discuss the factors governing the spectra of sulphates, whilst Adler and Kerr[403] and Schubert[528] have paid particular attention to variations in the spectra obtained from sulphate minerals. The infrared absorption spectra and interpretations for several metallic sulphates have been undertaken (Duval and Lecomte[149], Klimov[272], Steger and Schmidt[435] and Jaulmes[448]). Antikainen[498] has obtained spectra from the aqueous solutions of a large variety of the oxyanions of sulphur. A large number of sulphate reference spectra have been included by Nyquist and Kagel[4]. Haas and Sutherland[175] have made a detailed study of the crystal spectrum of gypsum. Thiosulphates have been investigated by Duval and Lecomte[212]. An infrared study of compounds of the type K_2XO_4 and $BaXO_4$ (where X = S, Se, Cr and Fe) has been made by Tarte and Nizet[437]. The far-infrared active lattice vibration modes of a number of sulphate minerals have been discussed by Karr[477], and the absorption bands of a number of sulphates in the far-infrared region have been given by Feairheller and Miller[610]. Rare earth sulphates (hydrated) have been investigated by Kharzeeva and Serebrennikov[574] and synthetic sulphate minerals by Uklonskaya[579].

SELENATES $(SeO_4)^{2-}$, TELLURATES $(TeO_4)^{2-}$ AND TANTALATES $(TaO_4)^{2-}$

Major absorptions are found in the regions 910–840 cm^{-1} and 460–390 cm^{-1} for selenates, and for tellurates in the regions 650–620 cm^{-1} and 360–290 cm^{-1}. Occasionally a weak shoulder in the 850–810 cm^{-1} range is found for selenates.

Several selenates have been examined by Duval and Lecomte[148], Sathianandan *et al.*[309] and by Caillere and Pobeguin[63]. The infrared spectra of some tellurates have been obtained by Siebert[285].

Rare earth orthotantalates have been studied recently by Petrov *et al.*[584].

PERIODATES $(IO_4)^-$ AND PERCHLORATES $(ClO_4)^-$

Hezel and Ross[415] included several perchlorates in their study of the infrared spectra of tetrahedral anions. Perchlorates absorb in the region 1150–1050 cm^{-1} and at 650–600 cm^{-1}, sometimes with a weak shoulder between 960 and 920 cm^{-1} probably due to ν_1 becoming active upon distortion of the tetrahedron. Periodates show an absorption band in the range 860–800 cm^{-1} with a weak shoulder on the low frequency side near 790 cm^{-1}, and two absorption bands between 340 and 225 cm^{-1}. However, very few compounds of this type have been studied and the full range of absorption may not be known. Ross[476] has shown that, for perchlorates, ν_1 is invariably present and that the degenerate ν_3 and ν_4 frequencies are often split, suggesting distortion of the $(ClO_4)^-$ tetrahedron.

OXYANIONS OF THE TRANSITION METALS: TUNGSTATES$(WO_4)^{2-}$, MOLYBDATES$(MoO_4)^{2-}$, PERRHENATES$(ReO_4)^-$, PERTENCHATES$(TcO_4)^-$ AND PERMANGANATES$(MnO_4)^-$

The major absorption bands for compounds of this type are given in *Table 4*. The normally inactive stretch ν_1 often occurs as a weak shoulder on the high frequency side and the degenerate ν_3 and ν_4 absorption bands are frequently split.

Table 4 INFRARED ABSORPTION BANDS FOR TRANSITION METAL OXYANIONS

	ν_3	ν_4
Tungstates $(WO_4)^{2-}$	900–750 cm^{-1}	420–250 cm^{-1}
Molybdates $(MoO_4)^{2-}$	970–740 cm^{-1}	400–250 cm^{-1} (frequently triplet)
Perrhenates $(ReO_4)^-$	940–880 cm^{-1} (multiplet) 3 ?	360–300 cm^{-1} (two or three peaks)
Pertechnates $(TcO_4)^-$	930–860 cm^{-1} (multiplet) 3 ?	350–315 cm^{-1} (two or three peaks)
Permanganates $(MnO_4)^-$	940–850 cm^{-1} (multiplet)	440–350 cm^{-1} (multiplet)

A study of the bending vibrations in the infrared spectra of tungstates has been made by Pudovkina *et al.*[305]. Scheelite- and wolframite-type tungstates have been recorded by Zorina and Syritso[572], and scheelite tungstates and molybdates by Tarte and Liegeois-Duyckaerts[580]. Ortho, meta, and paratungstates have been investigated by Dupuis[337]. Both anhydrous molybdates and tungstates were measured by Clark and Doyle[423] who showed that whilst isostructural compounds had similar infrared spectra, sufficient individuality existed to permit identification. Tungstates with the scheelite structure showed no ν_1 band. Alkali metal and scandium tungstates and molybdates were studied by Balashov *et al.*[546]. Tungstate and molybdate minerals have been the subject of an investigation by Yushkin and Bushueva[564].

Three vibration modes for the scheelite group of minerals are expected in the far-infrared region and these were observed by Khanna and Lippincott[419]. Busey and Keller[523] have studied molybdates, tungstates, perrhenates and pertechnates. Anhydrous permanganates, pertechnates and perrhenates have been investigated by Muller and Krebs[396], and perrhenates by Johnson, Rogers and Leroi[558]. Molybdates and chromates of some transition metals have been studied by Cord *et al.*[573]. Alkali–metal and ammonium pertechnates have been investigated by Petrov *et al.*[577].

OXYANIONS OF THE TRANSITION METALS: CHROMATES$(CrO_4)^{2-}$ AND FERRATES$(FeO_4)^{2-}$

Tarte and Nizet[437] have included chromates and ferrates in their studies. Duval and Lecomte[151], Dupuis[232] and Muller *et al.*[443] have all investigated the infrared spectra of chromates. Sufficient distortion of the chromate ion tetrahedron can give dichromate-like spectra, but the major absorption ν_3 occurs in the region 950–800 cm^{-1} with the ν_1 mode as a weak sharp peak or shoulder on the low frequency side. A doublet ν_4 is often observed in the 499–350 cm^{-1} region, whilst the ferrate absorptions occur in similar but somewhat lower frequency regions with absorptions between 900 and 750 cm^{-1} and 400 and 300 cm^{-1}. Nyquist and Kagel[4] have included chromate and dichromate spectra and the latter have also been recorded by Stammreich *et al.*[167]. Absorption bands of chromates and dichromates of a few elements in the far-infrared region are given by Feairheller and Miller[610].

GERMANATES $(GeO_4)^{2-}$

Germanates absorb at lower frequencies than the corresponding silicates, the isolated tetrahedron giving rise to absorption bands in the 800–700 cm^{-1} and 500–300 cm^{-1} region, depending upon the structure involved. Comparisons between germanates and silicates of similar structures have been made by Tarte and co-workers[532], spinels[44], ortho[75], soro[652], olivines and monticellites[121], and by Lazarev and Tenisheva[79] (chains). Dachille and Roy[499] have studied the olivine–spinel transitions in germanates and silicates.

SILICATES $(SiO_4)^{2-}$

The silicate group has been deliberately left until last since it deserves more emphasis than any other one group, due to the importance and abundance of minerals within this class. Silicates can be conveniently divided into four sections; classification within these four sections has been attempted with the aid of Winchell and Winchell[372] and Deer, Howie and Zussman[383]:

Ortho and ring silicates
Chain silicates
Sheet silicates
Framework silicates

Launer[35], Moenke[119], Saksena[301] and Prost[534] are among those who have summarised the regularities found in the absorption spectra of the various classes of silicate minerals. A detailed investigation of the infrared spectra of silicates has been made by Lyon[1] and more recently by Lazarev[500]. Matossi *et al.*[17,18] recorded the reflection spectra from a number of silicates in the late 1930's. All silicates absorb strongly in the range 1100–800 cm^{-1}, although the environment of the silicate ion as it varies from one mineral group to another affects the shape and frequency of the absorption bands. Thus, careful investigation of the spectrum of a silicate mineral can place that mineral in its correct classification.

Ortho and Ring Silicates (Neso-, Soro- and Cyclo-)

This group is taken to include nesosilicates (single SiO_4 tetrahedra), sorosilicates (groups of two or more SiO_4 tetrahedra, usually Si_2O_7) and cyclosilicates (rings of tetrahedra). Tarte has been responsible for several studies into the spectra of silicate minerals within this group—garnets[43], orthosilicates (compared with orthogermanates)[75] and olivine and monticellite structures having been investigated[121].

NESOSILICATES

Olivine group Minerals of this group exhibit complete diodochy between atomic pairs, e.g. Mg and Fe^{II}, Fe^{II} and Mn, etc., i.e. between fosterite–fayalite and between fayalite–tephroite, etc. Only the fosterite–fayalite, but that in some detail, is shown by Lyon[1], whereas Duke and Stephens[281] have made a detailed study of the olivine group, including the fosterite–fayalite, fosterite–monticellite, monticellite–glauchroite, glauchroite–tephroite and the fayalite–tephroite series. They showed that within any one substitution system, the frequency of each absorption band varied systematically with the ratio of the substituent cations. Four absorptions were found in the 1000–800 cm^{-1} region. The absorptions near 975 cm^{-1} and 875 cm^{-1} are large and fairly broad with a weak shoulder between them, and a sharp peak on the low frequency side. A fifth peak arises between 625 and 550 cm^{-1}. The distinction between sillimanite and mullite has been the subject of papers by Roy and Francis[264], Tarte[334] and Hariya *et al.*[459]. The effects of compositional changes on the spectra of olivines has been the subject of a study by Burns and Huggins[569] and isotopic substitution is used as an aid to assignments by Paques-Ledent and Tarte[651].

Garnets Two series are known 'pyralspites' (*py*rope, *al*mandine, *sp*essartite) and 'ugrandites' (*u*varovite, *gr*ossularite, *and*radite), and intermediate compositions within each group, but not between the groups, commonly occur. The general appearance of the spectrum is similar for all of the members of any one group, but absorption frequencies vary with cation substitution and lattice parameter. Tarte[43] has studied both the pyralspite and ugrandite series, as has Moenke[202], and more recently the garnets have been studied by Moore and White[556,626] and Valueva[559]. Wickersheim *et al.*[183] have paid particular attention to rare earth iron garnets.

Garnet spectra are characteristic and differ from the olivines in the disappearance of the sharp absorption peak found on the low frequency side of the major bands near 975–875 cm^{-1}. Clark[483] has obtained the near-infrared spectrum of single crystals of three garnets (almandine, pyrope and grossularite).

SOROSILICATES

Being more complex structurally, the sorosilicates indicate this by their infrared spectra. A study of silicates with anions/groups of the type Si_2O_7 has been undertaken by Lazarev[52,80], who showed that there are absorptions in the 1100–800 cm^{-1} region, and below 500 cm^{-1}. The additional band between 680 and 650 cm^{-1} distinguishes sorosilicates from nesosilicates. Strens[250] and Sitzia[637] showed a shift in the 1100–800 cm^{-1} absorption band of epidotes with a change in the iron content. Further studies have been made by Andreenko and Plyusnina[526] and both silicates and germanates with Si_2O_7-type groups have been studied by Tarte *et al.*[652].

CYCLOSILICATES

Trigonal, tetragonal and hexagonal rings are known, but no specific reference to tetragonal ring silicates has been found. Considerable effort has been made by Plyusnina *et al.*[50,364] in an attempt to relate the absorption spectra of various cyclosilicates with structure. They found that six-membered rings absorbed in the 800–770 cm^{-1} region, whilst three-membered rings absorbed in the 770–740 cm^{-1} region.

A study of beryl (and other beryllium minerals) has also been made by Plyusnina[172,325] and by Povarennykh and Nefedov[571]. Lyon[1] has shown that compositional changes in beryls may be clearly defined in the 1060–1000 cm^{-1} region, and that cordierite spectra, although comparable to the spectra of beryls, differ in the 800–700 cm^{-1} region in having fewer peaks.

Chain Silicates (Ino-)

Single, double, triple and quadruple chain silicates constitute this group. Triple chains (eudidymite and epididymite) are not discussed but the spectra are included in Part III. Quadruple chains, i.e. the epidote group, have been included in the sorosilicate section owing to the presence of Si_2O_7 groups in the mineral. The absorption spectra of pyroxenes and other chain silicates (and germanates) have been studied by Lazarev and Tenisheva[78,79].

SINGLE CHAINS

Orthopyroxenes Lyon[1] has included the detailed absorption spectra of 21 orthopyroxenes, which indicate the spectral changes with substitution of Mg and Fe in the enstatite–hypersthene series.

Clinopyroxenes A number of clinopyroxene inosilicates have been given by Lyon[1], who has shown that the spectra from diopside and hedenbergite are clearly different from other pyroxenes. Examination of the 775–530 cm^{-1} region by Ryall and Threadgold[385] indicated that the number of peaks gave an indication as to the number of tetrahedra in the repeat unit of the chains. Other spectra–structure correlations have been made by Rutstein and White[467].

DOUBLE CHAINS

Amphiboles Because of the complexities introduced by the possibility of considerable substitution within these chains, the spectra of amphiboles are very difficult to interpret. A number of amphiboles are included in Part III but no correlations are attempted. Beiseev and Kiselev[529] have studied amphibole asbestos.

Sheet Silicates (Phyllo-)

Layer lattice groups have been extensively studied by Stubican and Roy[88,95], Beutelspacher[153] and by Farmer alone[327] and with Russell[370], and in the far-infrared by Ishii *et al.*[525]. Their studies have indicated that substitutions are usually readily detected and that a mineral may be placed in its proper position in an isomorphous series without recourse to chemical analysis. The above-mentioned papers are extremely useful to those particularly interested in the infrared analysis of phyllosilicates. Glauconites and talcs have been studied in particular by Manghnani and Hower[312] and Wilkins and Ito[501], respectively. Brindley and Zussman[99] and Luce[481] have investigated the infrared spectra of the serpentines, as have Ptak and Pampuch[536], who have also studied 1 : 1-type laminar silicates[539]. A number of mica varieties are included in studies by Lyon and Tuddenham[42], Melchakova and Plyusnina[560], Tarasevich and Sivalov[644], and Vedder[251] has suggested correlations between the infrared spectrum and chemical composition of micas. Kireev[547] has studied the isomorphous substitution of tetrahedral Al in micas, and likewise in phlogopites by Zorina[641].

Considerable attention to the vibrational modes of the hydroxyl group has been given, particularly by Farmer and Russell[370], Wilkins and Ito[501], Vedder[251] and by Serratosa and Bradley[317]. Chemically analysed natural biotites have been recorded and variations in absorption intensities in the 770–670 cm^{-1} region have been related, although positive assignment was not possible, to the corresponding variations in the calculated tetrahedrally co-ordinated aluminium, by Liese[444]. Tuddenham and Lyon[168] and Hayashi and Oinuma[485] have shown that by relating infrared spectra to the chemical analysis of chlorites and related minerals, both the degree of substitution of Al for Si (the Y number) and the total iron content of the mineral, can be estimated from the infrared data. For other phyllosilicates, Lyon[1] may be referred to. The wealth of information that now abounds on perhaps the most well-studied group of silicate minerals, is by far too detailed and specific for inclusion within these pages and the reader is referred to the works of his choice mentioned above.

The vibrational modes of some phyllosilicates observable in the far-infrared region of the spectrum have been reviewed by Karr[477].

CLAY MINERALS

Recent infrared studies have all shown the usefulness of the infrared technique for identification not only of the species, but also characteristics of the constituent minerals that would hardly be identified by the X-ray technique. Studies in the OH absorption region of clay minerals have been made by Scholze and Dietzel[140], Wolff[291] and Touillaux *et al.*[311], with the conclusion that even closely related minerals can be distinguished in this region if an instrument of sufficient resolving power is used. A useful review paper is that by Fripiat[316] who includes details regarding the identification of absorbed species on mineral surfaces. Other studies include those by Farmer and Russell[551–554] and by Tarasevich[668]. Kaolinites have been the particular interest of Blasezak and Pampuch[622] and Reh[656], and a method of classification of kaolinites by infrared spectroscopy has been devised by Parker[653]. Studies of water on clay have been made by Prost[654,655], and the diffuse reflection spectra of the OH region have been made by Lindberg and Snyder[555]. The influence of controlled cationic substitution on the infrared absorption spectra of various clay minerals has been studied by Stubican and Roy[503] and Tettenhorst[504]. Cation migration in clays has been investigated by Calvet and Prost[550]. Torok used infrared spectroscopy to study structure changes of illite and montmorillonite[657]. Recently, the far-infrared spectra of clays has been obtained by Larson, Pardoe and Gebbie[633].

Framework Silicates (Tecto-)

SILICA GROUP MINERALS

Quartz (SiO_2) and its polymorphs, although oxides by composition, are placed as a group in the tectosilicate subclass of the silicate minerals, since its crystal structure consists of a three-dimensional framework of SiO_4 tetrahedra. Haccuria[253] alone and with others[310] has carried out an investigation into the infrared spectra of various silica polymorphs in order to determine the spectral differences between α-quartz, cristobalite, tridymite, vitreous and amorphous silica. Reitzel[263] (α-quartz, cristobalite and vitreous silica in the 600–400 cm^{-1} region), Lippincott *et al.*[304] (coesite, cristobalite, tridymite and vitreous silica) and Sclar, Carrison and Schwartz[457] (α-quartz, coesite and stishovite) have all been involved in investigation of minerals within the group. Coesite (Austin *et al.*[294]), tridymite (Sun[105]) and geyserite (Shimida and Komaki[367]) have been the subject of individual investigations. Reflection spectra of α-quartz, cristobalite and vitreous silica have been obtained by Simon and McMahon[464].

Distinction between the various forms of silica is possible. Fused quartz and α-quartz give similar spectra, except that in the latter case the absorption band at 800 cm^{-1} partly resolves into a doublet at 798 and 779 cm^{-1}. Tridymite and cristobalite, whilst having similar absorption bands to α-quartz in the 1100 cm^{-1} region, differ in giving rise to absorptions at 568 cm^{-1} and 621 cm^{-1}, respectively, where α-quartz has none. Coesite and stishovite have highly characteristic spectra and are easily distinguished. The low frequency absorption bands of some of the silica polymorphs have been discussed by Karr[477], and these again allow distinction between members of the group.

FELDSPARS

Plagioclase feldspars varying in composition from 100% anorthite, through bytownite, labradorite, andesine and oligoclase to 100% albite, show variations in the intensity, shape and position of the absorption bands. These variations have been noted by Lyon[1], Thompson and Wadsworth[282] and Hafner and Laves[340], particularly in the 650–500 cm^{-1}

region, and by Angino[486] in the far-infrared region. The infrared spectra of some 57 tecto-silicates, along with some correlations and interpretations of the various absorption bands, have been presented by Milkey[111], and by Bahat[537] and Liese[540]. Synthetic zeolites have been studied by Shikunov *et al.*[567] and many natural zeolites were studied by Belitskii and Golubova[582]. A number of scapolites have been studied by Wehrenberg[629].

PART III

Explanatory Notes and Data Tables

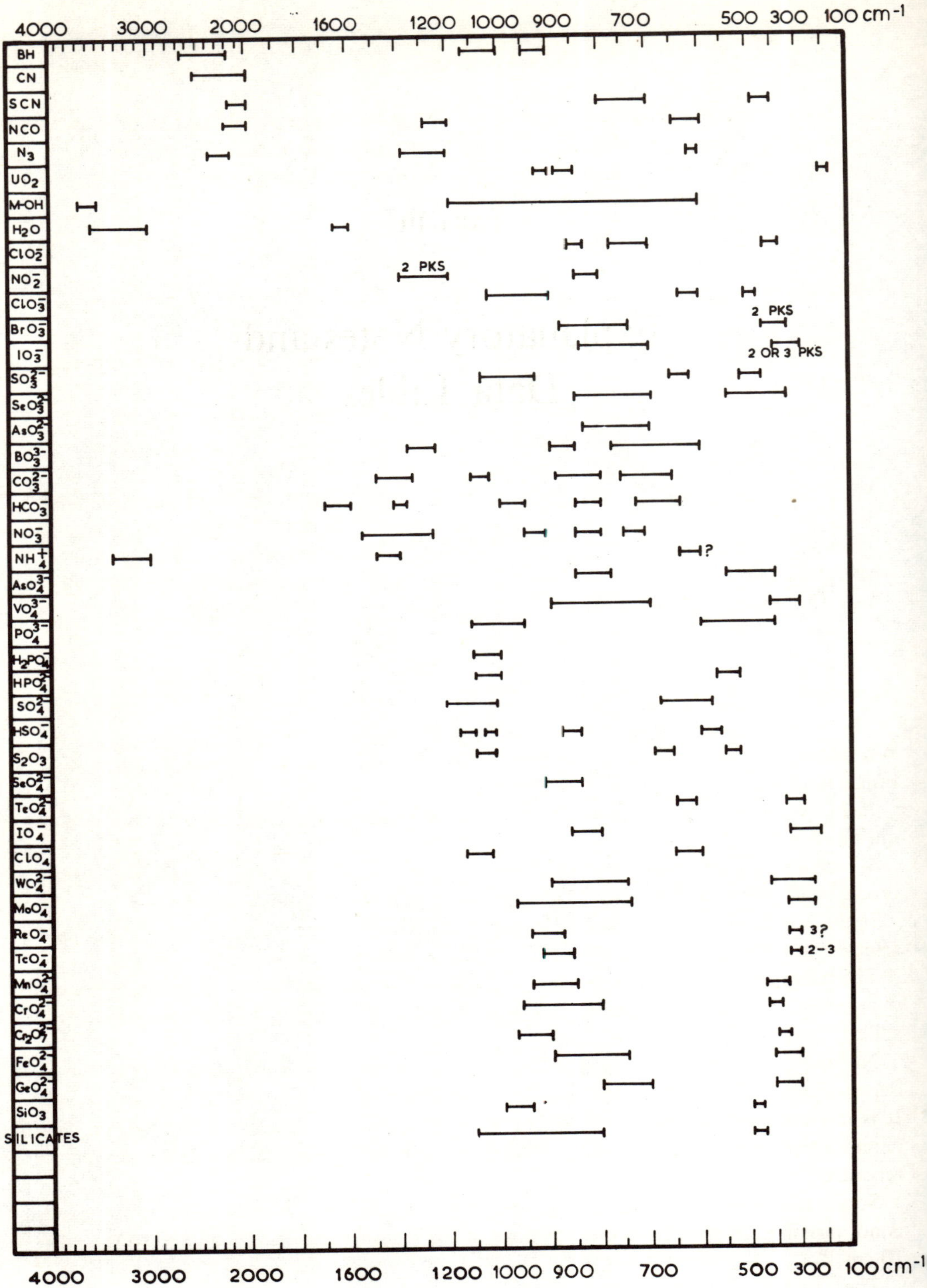

Figure 6. Inorganic group frequency correlations

EXPLANATORY NOTES

Part III contains spectral data of minerals and other related inorganic compounds, collected from the literature, tabulated together with approximate relative intensities and some band shape information. The absorption band tables express the frequency in wave numbers of maximum absorption (cm^{-1}).

Abbreviations

vw	very weak	$< 10\%$ absorption
w	weak	$< 20\%$ absorption
m	medium	$< 40\%$ absorption
s	strong	$< 80\%$ absorption
vs	very strong	$> 80\%$ absorption
var	variable intensity	
b	broad	
vb	very broad	
sp	sharp	
inf	inflection (change in slope)	

w,sh	weak shoulder—no appreciable resolution
sh	shoulder—just resolved
m,sh	medium shoulder—quite distinct shoulder
s,sh	strong shoulder—approaching in intensity to the main absorption band
plat	plateau—flat top to absorption band or shoulder

Numerical data

Where two or more spectra are known, figures quoted are either corroborated or else the range is given.

If an absorption band is in doubt or is uncorroborated then the data are followed by a question mark (?).

When a series of absorption bands are closely spaced they are grouped together as a broad band, with an indication of the number of partly resolved bands.

If the bands are of approximately equal intensity, but are only just resolved, they are referred to as being either a doublet or triplet, which may or may not be broad.

The range of values quoted does not distinguish between instrumental errors, preparative differences, and band frequency shifts due to compositional differences (isomorphous shifts).

No indication of sample preparation is given. The absorption band positions may vary depending on whether discs or mulls are used, in which case the absorption range is given. It is not expected that identification of unknowns will be made from the data contained herein alone, but that either the original literature will be sought or a reference spectrum recorded.

Some spectra are presented in such a way that calculation of accurate absorption maxima is not possible. In such cases the positions of the maxima are usually quoted to the nearest 10 cm^{-1}.

The absorption bands tabulated only indicate those bands that have been found. It is conceivable that the reader may find more or less bands in his spectrum, and must then decide whether impurities are present and on the quality of the identification based on the 'degree of fit'.

The data given are basically of named minerals, whether natural or synthetic. However, some related synthetic inorganic compounds are included either as examples of a type or for purposes of comparison. Synthetic minerals and inorganic compounds are denoted by the suffix (*S*).

References are given to those inorganic compounds and minerals where either the data are considered insufficiently confirmed or are of lesser interest to mineralogists.

The names of minerals and their chemical formulae have been given, using the books by Hey[671] and Fleischer[670] as references. In cases where two names exist for the same mineral both are quoted, and they are separated by an oblique stroke, e.g., SPHALERITE/BLENDE.

ELEMENTS

1
CARBON

Diamond Type I	& Type II	Graphite
1540sh		1165sh } b
1520m		1150sh } b
1426s, sh		1110w } b
1372inf		
1332vs	Absence of strong bands in this region (1332–1003)	No bands between 900 and 100 cm^{1}
1282sh		
1203vs		
1171sh		
1093b, sh		
1003		
780		
480		
328		
References *178, 203, 206* (Xtals), *588, 589*		*References* *3, 21*

2
SULPHUR

Rhombic	& Orthoclinic
1525s	
1310m	1175sh } b
1050w	1150sh } b
990w	1110w } b
935m	690vw
905sh	465vw
870sh	
845vs	
710w	
680m, sh	
References *170* (Xtals)	*References* *3, 21*

3
SILICON

No absorptions between 900 and 400 cm^{-1}

Reference 21

4
SELENIUM

Amorphous
1150vw, b
740m, b
ca. 485vs, b

Reference 217

5
GERMANIUM

870w	Absorption bands quoted are those given by Omori[3]
585vw	
549vw	
518vw	

References 3, 510

BINARY COMPOUNDS

6
BORIDES

No absorption bands are quoted, since any that do occur are usually broad and weak.

However, borides often have a broad medium-to-strong absorption near to 90 cm^{-1}.

The following borides are given by Nyquist and Kagel[4]:

Al, Mg, Ca, Si, V, Cr, Zr, Nb, Mo, La, Ta, W

References 4, 60

7
NITRIDES

The number of absorption bands depends on the number of atoms in the molecule.

A fairly comprehensive list of nitrides (and azides) is given by Nyquist and Kagel[4]:

Ca, Ba, B, Cr, Al, Si, Ti, V, Zr, Nb, Mo, Ta.

Zn, Li, Cu, Mg are given by Brame *et al.*[60]

References 4, 60

8
AZIDES

For the azides of NH_4, Na, K, Rb, Cs, Ba see Nyquist and Kagel[4].

Reference 4

9
SILICIDES

For the silicides of B, Ca, Mg, Mn, Mo, Nb, Ti, V, W, Zr see Nyquist and Kagel[4].

Reference 4

10
SELENIDES

The number of absorption bands depends on the number of atoms in the molecule.

The selenides of the following are given by Nyquist and Kagel[4]:

Ga, Sn, Pb, Ti, Cr, Nb, Mo, Ta, W, Zr, Zn, Cd

Mitsuishi *et al.*[200] quote absorption maxima for ZnSe and CdSe.

ZnSe 215 cm^{-1}
CdSe 185 cm^{-1}

References 4, 200

11
TELLURIDES

For the tellurides of Sn, Bi, Ti, V, Cr, Zn, Cd, W, Mo, see Nyquist and Kagel[4].

Mitsuishi *et al.*[200] quote absorption maxima for ZnTe and CdTe.

ZnTe 190 cm^{-1}
CdTe 150 cm^{-1}

Tetradymite (*Ref. 576*)

References 4, 200

SULPHIDES

12
REALGAR
4[As_4S_4]

373–70s	205m
367s, sh } 362sh	194m
358m } 362sh	183m
343vs } 337s	172–0s
330sh } 337s	148vw, sh
284w, sh	141vw, sh
269vw, sh	(109)?
258vw, sh	47w
225s	37vw
211m	

References 462, 576

13
ORPIMENT
4[As_2S_3]

393m, sh	(295)sh
381–79m	202m, sh
361m, sh	184s, sh
354–50m	160w, sh
348–45m, sh	139vs, b
310–05s, b	52w

References 462, 576 (other sulphides of arsenic *Ref. 4*)

14
ARSENOPYRITE
8[FeAsS]

1110w 435w

The spectrum is virtually free of absorption bands

References 3, 21, 576

15
CHALCOPYRITE
4[$CuFeS_2$]

1175sh }
1135sh } b
1110w }

References 3, 21

16
PYRITE
4[FeS_2]

1190sh }
1150sh } b Reported only by
1125sh } Omori[3]
1110w }
418s
412sh

References 3, 21, 33, 279, 576

17
MARCASITE
2[FeS_2]

1175s } b Reported only by
1110w } Omori[3]
480sh
422s

References 3, 21, 33, 279, 576

18
CINNABAR
3[HgS]

350–340s } doublet?
285–276m–s }
130–123m–s, vb
89–87m, sh
38w

References 4, 7, 11, 32, 576

19
SPHALERITE/BLENDE
4[ZnS]

1175sh	472vw, b
1110w	417vw, b
770b, sh	310vs, b*
532vw, b	*Most important band

References 3, 4, 21, 200, 576

20
BOURNONITE
$4[PbCuSbS_3]$

1110w, b
620w
The spectrum is virtually free of absorption bands

References 3, 21

21
GALENA
4[PbS]

1175sh }
1160sh } b
1100w }
No other absorption bands are to be found above 400 cm^{-1}

References 3, 21, 33, 576

OTHER SULPHIDES

		References
Stibnite	$4[Sb_2S_3]$	*4, 11, 576*
Barium 180 cm^{-1}	BaS	*11*
Bismuth	Bi_2S_3	*4, 11*
Calcium 255 cm^{-1}	CaS	*11*
Cadmium 270–60 cm^{-1}	CdS	*4, 200*
Copper	Cu_2S	*4*
Molybdenite	$2[MoS_2]$	*4, 576*
Nickel	NiS	*4*
Niobium	NbS	*4*

		References
Silver	Ag_2S	*4*
Strontium 208 cm^{-1}	SrS	*11*
Tantalum	TaS & TaS_2	*4*
Tellurium	TeS_2	*4*
Titanium	Ti_2S_3	*4*
Tungsten	WS_2	*4*
Pyrrhotine	2[FeS]	*576*
Wurtzite	2[ZnS]	*576*
Heazlewoodite	$[Ni_3S_2]$	*576*
Hauerite	4[MnS]	*606*

OXIDES

22

CORUNDUM

$2[\alpha\text{-}Al_2O_3]$

(800)b, sh
790–750vs
ca. 720vs
640m, sh
600–575vs

(Four peaks occur between 790 and 560 cm^{-1} giving a broad absorption band: 790–750vs, ca. 720vs, 640m, sh, 600–575vs)

520vw
490sh
450–32vs
375sh

References 4, 60, 84, 416, 450, 451*, 508, 512** (*Xtals)

23

γ-ALUMINA

$\gamma\text{-}Al_2O_3$

1180vw
825–20vs, b
752–45vs, b
662
580
423vw

(825–20vs, b and 752–45vs, b: b doublet; 662 and 580: plat; 1180vw to 423vw: vb)

References 1, 495

24

ARSENOLITE CLAUDETITE

$16[As_2O_3]$ $4[As_2O_3]$

Arsenolite	Claudetite
1020	
830sh	837m
795s	641vs, b
750sh	553s, b
500–490sh	465m, b
477–5s, b	
355sh	
340s	

References 7, 32, 137, 279, 416, 619

25

ARSENIC PENTOXIDE

As_2O_5

ca. 810s, b
575w
530w
420m, sh
395m, sh
340s

Reference 32

26

SENARMONTITE VALENTINITE

$16[Sb_2O_3]$ $4[Sb_2O_3]$

Senarmontite	Valentinite
740vs	
685sh	690s
590w	590s
550w, sh	550s, vb
482w, b	

383s
345w
322w
265s, b
180w
} Synthetic Sb_2O_3

References 4, 7, 32, 279, 332, 416, 619

27

ANTIMONY PENTOXIDE

Sb_2O_5

950vw
ca. 740vs, b
685–0b, sh
450m, sh
370sh

References 4, 7, 32

28

BROMELLITE (*S*)

2[BeO]

ca. 1700b, sh
1600m, vb
ca. 1400vs
1350sh
1100–600vs, vb
No peaks between
600 and 250 cm^{-1}

References 379 (Xtals), *416*

29

CHROMIUM OXIDE (*S*)

Cr_2O_3

1100w
710inf
650vs
570b, sh
555vs
} b doublet
440–35w, sp
407m, sp
308w

References 4, 60, 344, 416, 439, 512 (Xtals), *513*

30

COBALT OXIDE (*S*)

Co_3O_4

670vs
655sh
635sh
588vs
575vs
} b doublet
550m, sh
460b, sh
Far-I.R. Nyquist and Kagel[4]

References 4, 416, 495

31

CUPRITE (*S*)

$2[Cu_2O]$

1124vw
794sh
699inf
662sh
625–10vs* *Major band
518sh
430sh
148s

References 4, 9, 11, 85, 416, 510, 595

32
HEMATITE
$2[Fe_2O_3]$

1175sh, 1160sh, 1150w, b — b — Reported only by Omori[3]

560–50vs, b
532b, sh
474–67s, b
370sh
350inf
310vs, b

References 3, 4, 9, 21, 174, 416, 451, 508, 533 (Xtal-reflection), *601, 363* and *635* (both include Maghemite)

33
MAGNETITE
$8[Fe_3O_4]$

575–0s, b
410–380s, vb (lower frequency preferred)

References 3, 4, 9, 174, 407, 416, 508, 601 (also Wustite $Fe_{<1}O$), *635*, (Magnetite → Maghemite → Hematite)

34
PYROLUSITE
$2[MnO_2]$

685–10vs, vb (lower frequency preferred)
550–35b, sh
420–00m, b, sh
340–20m, b, sh

References 3, 4, 9, 279, 416, 452 (poor quality), *606*

35
RAMSDELLITE
$4[MnO_2]$

700–680b, sh, 600–500vs, vb, 498–470s, vb — vb

Reference 279

36
HAUSMANNITE
$8[Mn_3O_4]$

1150m, 1105sh, 1087m, 1050sh, 1015sh — b — Reported only by Omori[3]

728sh, 602vs, b, 525–20s, vb, 475–60s, b — vb

418s, b

References 3, 9, 21, 116, 416, 606

37
CRYPTOMELANE
$[KMn^{II}\ Mn^{IV}{}_8\ O_{16}\ .\ H_2O]$

ca. 710s, vb
600–580vs, ca. 525vs — b doublet

References 3, 452 (both poor spectra)

38

BRAUNITE

$8[Mn^{II}, Mn^{III}_6 SiO_{12}]$

ca. 1112vw doublet?
1040vw
1015vw
953s, b
850b, sh
815b, sh
712m, sh ⎫
625vs, vb ⎪
555inf ⎬ vb
530vs, vb ⎪
480s, sh ⎪
455sh ⎭

References 3, 9, 116, 606

39

MASSICOT

$4[PbO]$

Massicot
505–00vs, b
377s
300s

LITHARGE

$2[PbO]$

Litharge
470–65s

References 279, 416

40

α-QUARTZ

$3[SiO_2]$

(Chalcedony, Chert and Novaculite)

ca. 1200vw, sh
1172–70sh
1150sh
1090–76vs, b
915vw
805–796m ⎫ b doublet
785–775m ⎭
697–2w, sp
668vw
522–03m, sh
479
462–50s
390–60m doublet?
257w

References 3, 9, 21, 33, 153, 255, 263, 294, 304, 310, 416, 457, 464*, 465, 592, 508*
(*Reflection spectra)

41

TRIDYMITE

$64[SiO_2]$

1175b, sh ⎫ b
1109–5vs, b ⎭
794–87m
570–68w
508sh
478vs, b
392–85m
ca. 333

References 1, 105, 255, 304, 310, 508

42
α-CHRISTOBALITE
$4[SiO_2]$

1210–00m, sh
1160sh
1100–1090vs
800–792m
725b, sh
623–17m
520–15sh
490–82s, b
372m
300–295m

References 4, 9, 255, 263, 294, 304, 310, 464*, 508.*
(*Reflection spectra)

43
COESITE
SiO_2

1225–18s, sp, sh } b
1170–52b, sh } b
1098–77vs, b } b
1040–36m, sp, sh } b
837sh
814sh
796–4m, sp
683s, sp
600–598s, sp } doublet
561–57s, sp } doublet
490m, sp, sh
448–42sh
435–30s, sp
390sh
340w

References 1, 294, 304, 457

44
STISHOVITE
$(SiO_2)_n$

949b, sh } vb
885vs } vb
769sh } vb
730m, b, sh } vb
672vs } vb
628s, sh } vb
560s } vb

References 1, 298, 457

45
OPAL
$SiO_2 . nH_2O$
(Diatomite, Tripolite and Geyserite)

3422s, b
1622w
ca. 1220inf
1102–1090vs, b
790–85s
476–70s

References 1, 9, 33, 105, 367 (Geyserite), *613*

OTHER FORMS OF SILICA

	References
Fused:	*351*
Silica gel:	*31, 33, 162, 416*
Glass:	*4*

Lechatelierite (vitreous)

References 263, 304, 310, 464, 514

46
PERLITE
$SiO_2 \cdot nH_2O$ Var. Opal

ca. 3500
ca. 1150sh
ca. 1050vs, b
788w
765sh
695b, sh
632vw
584vw
453s
426b, sh

References 1, 211

47
CASSITERITE
$2[SnO_2]$

700sh
640s, b
540m, vb, sh
440vw

References 3, 4, 9

48
RUTILE
$2[TiO_2]$

695s, b } centre at
525s, vb } 625–608s, b
423–15s
(390)sh
352s
320sh

References 4, 9, 21, 66, 279, 416, 451, 465, 612

49
ANATASE
$4[TiO_2]$

700s, vb
525–460 centre vs, vb peak
347m

References 9, 279, 416

50
BROOKITE
$8[TiO_2]$

680b, sh
535s, vb
410

Reference 9

51
ZINCITE
$2[ZnO]$

615vb, sh
535b, sh
440m, vb

References 3, 4, 9, 416

OTHER OXIDES

		References
Barium	BaO, BaO_2	*332, 416*
Bismuth	Bi_2O_3	*416, 619*
Boron	B_2O_3	*145, 511, 416*
Calcium	CaO	*4, 332, 416*
Cerium	CeO_2	*4, 255, 416*
Dysprosium	Dy_2O_3	*4, 382, 416*
Erbium	Er_2O_3	*4, 382, 416*
Gadolinium	Gd_2O_3	*382, 416*
Gallium	Ga_2O_3	*145, 416*
Germanium	GeO_2	*4, 304, 310, 416*
Hafnium	HfO_2	*4, 416*
Holmium	Ho_2O_3	*4, 382, 416*
Indium	In_2O_3	*4*
Iodine	I_2O_5	*4*
Lanthanum	La_2O_3	*390, 416*
Lead	Pb_3O_4	*4*
Lutetium	Lu_2O_3	*382, 416*
Magnesium	MgO	*4, 33, 388, 416, 450*
Manganese	MnO	*4, 388*
	Mn_2O_3	*332, 451*
Mercury	HgO	*4, 416*
Molybdenum	MoO_3	*4, 332, 516*

		References
Neodymium	Nd_2O_3	*416, 510*
Nickel	Ni_2O_3	*4, 332, 416*
Niobium	NbO_2, Nb_2O_5	*4, 416*
Samarium	Sm_2O_3	*4, 416*
Scandium	Sc_2O_3	*390, 416*
Selenium	SeO	*271*
Silver	Ag_2O	*4*
Strontium	SrO	*332, 416*
Tantalum	Ta_2O_3	*4, 416*
Tenorite	4[CuO]	*4, 332, 416, 451, 595*
Thallium	Tl_2O_3	*4*
Thorium	ThO_2	*4, 255, 416*
Thullium	Tm_2O_3	*382, 416*
Tungsten	WO_3	*4*
Uranium	UO_2, U_3O_8	*4, 138* UO_2 also
Vanadium	V_2O_5, V_6O_{13}	*4, 416*
Ytterbium	Yb_2O_3	*4, 382, 416*
Yttrium	Y_2O_3	*4, 382, 390, 416*
Zirconium	ZrO_2	*4, 415*
	ZrO_2, xH_2O	*321*

MINERALS	*Reference*
Todorokite (Mn, Ba, Ca, Mg) $Mn_3^{IV}O_7 . H_2O$	*3*
Hollandite nr[(Ba, Na, K)Mn^{II}_2 $Mn^{IV}_{6.5}$ $O_{16} . H_2O$]	*3*

52
ATACAMITE
$4[Cu_2(OH)_3Cl]$

3440
3340
3220
1650w, b
1390vw, sp
1165vw, b
987m, 952m, 917m, sh, 897m, 866m, 852m, 822m, sh — 7 sharp medium peaks
598m, b, 515m, 483m, 445m, b, 410? — b

References 9, 116, 618

53
BISCHOFITE (*S*)
$2[MgCl_2 . 6H_2O]$

3420s, vb
3235s, vb
1630vs
720sh
612s, b
470s, vb

References 4, 9, 608

54
CARNALLITE
$12[KMgCl_3 . 6H_2O]$

3415–00vs, vb
3240–35sh
1640s
ca. 1110w, vb
720–10sh
620–10m, b
480–65s, b
220m, sh
170s
145sh
120b, sh

References 4, 9, 116, 608

55
FLUORITE (*S*)
$4[CaF_2]$

Major absorption band is at
410inf
275vs, b

References 4, 9

56
HALITE (*S*)
$4[NaCl]$

The absorption band begins at ca. 650 cm^{-1}, with total absorption occurring at ca. 400 cm^{-1}.

References 4, 9

57
KOENENITE
$Mg_5Al_2Cl_4(OH)_{12} . 2H_2O$

1640s
1410vw
1170w
1130sh
860vb, sh
690vs, vb
560m, b, sh
475sh
452s

Reference 9

58
RINNEITE
$2[K_3NaFeCl_6]$

1550m, sh
1520m
ca. 540vb

References 9, 608

59
MALLADRITE (*S*)
Na_2SiF_6

740s, vb
525w }
497s } triplet
477s }

References 7, 32

60
CRYOLITE
$2[Na_3AlF_6]$

630sh }
605sh } vb
585vs }
ca. 410?

References 3, 9

61
SYLVITE (*S*)
$4[KCl]$

215m, sh
145vs, b

References 4, 9

62

NOCERITE

$3[Ca_3Mg_3O_2F_8]$

1270s, b
1115vw
910w, b
862sh
740m
625b, sh }
480s, b, sh } vb
412vs }

Reference 10

63

SAL AMMONIAC (*S*)

$4[(NH_4)Cl]$

3200sh
3130s }
3080 } b doublet
3040–30s }
2800m, sh
1445sh
1405m
720vw
475vw
220b, sh
170vs

References 4, 9

HYDROXIDES

64
BAYERITE (*S*)
2[$Al(OH)_3$]

1021s	b doublet
974s	b doublet
860sh	vb
764vs, b	vb
625b, sh	vb
585m, sh	vb
555sh	vb
525vs	vb
458sh	
425vw	

References 1, 318, 292, transformation into Hydrargillite

65
BOEHMITE
4[AlO . OH]

3290–45s, sp	
3095–40s, sp	
(2780)	
1630w	
1160–35b, sh	
1085–65s	
760–30vs	b triplet
630–05vs	b triplet
525–00vs	b triplet
410–00	
372–0	
325	

References 1, 4, 9, 127, 295, 279, 318, 321, 357, 362

66
DIASPORE
4[AlO . OH]

2920–2895vs, b
2325
2115–05m, sp
1985–0m, sp
1635w
1085–70vs } b
970–60vs } b
756m, sh
720–05s
680s, vb, sh
580–70m–s, vb
521–15m, sh
375
350

References 3, 9, 127, 279, 318, 357, 362, 515

67
GIBBSITE/HYDRARGILLITE
8[$Al(OH)_3$]

3685
3617s, sp
3520vs } b
3445–28vs } b
3390–60s, sh } b
1630–00w
1030–18vs
980–58m, sh
920–10w, sh
800s } b doublet
749–40s } b doublet
670–60s, sh } vb
585m, sh } vb
562–40vs, b doublet? } vb
515vs } vb
(476)
455–0m, sp
425–0 } w, b doublet
410 } w, b doublet
370s
285m
255m
196w

References 1, 4, 9, 81, 127, 153, 318, 516, transformation to Corundum

68
NORDSTRANDITE (*S*)
$Al(OH)_3$

3655s, sp
3625sh
3565s, sp, sh
3545s } b doublet
3500s } b doublet
3430s } b doublet
3375s } b doublet
1060s, sh } b
1035sh } b
1010sh } b
965vs } b
916s, sh } b
870sh
830s, b
ca. 760b, sh
ca. 710vs, vb

Reference 356

69
BEHOITE
$Be(OH)_2$

3545vs
3510sh
3470sh
1075vs, b
877m, sh
782vs } b doublet
752vs } b doublet
720sh

Reference 455

70
PORTLANDITE
[$Ca(OH)_2$]

3640s, sp
3530sh
ca. 3400sh
1600w
ca. 1430m, b doublet
1090–50w, b
ca. 870w
650m
450
370
320

References 46 (OH region), *193*, *321*

71
BRUCITE
[$Mg(OH)_2$]

4290w
4050m
3730s
3570w
3290m
2900w
2525w
1630w, b
690–40b, sh
575–50b, sh
465–30vs, vb
370s, sh

References 3, *4*, *46**, *49**, *157** (Xtals), *193*, *314*, *321*, *347** (Xtals), *585* (*OH region)

72
GOETHITE
4[FeO . OH]

3095–2985m, vb
ca. 1660
1105m, b
1040b, sh
912–882vs doublet?
812–793vs doublet?
672m, sh } vb
599–578s, b } vb
470–450s } vb
410sh } vb

References 9, *70*, *116*, *127*, *279*, *454*, *515*

73
AKAGANEITE (*S*)
8[β-FeOOH]

3450sh
3350vs, b
1630w, b
1520w
1420b, sh
1340m
1295sh
845s, b, sh
675vs, b
450vs, vb
275s, b, sh

Reference 4

74
LEPIDOCROCITE
$4[\gamma\text{-FeOOH}]$

3390
3125
2857m, b
1165–45w, b
1028–12s, sp
887w
790sh
750–38m
520inf
480–65vs, vb
410?

References 9, 70, 116, 279, 515, 636 (Lepidocrocite → Maghemite → Hematite)

75
HETEROGENITE/STAINIERITE
(*S*)
CoO(OH)

3400w, vb
ca. 1960m, vb } vb
1640m } vb
1205sh } vb
662m
598vs

Reference 125

76
SASSOLITE
$4[B(OH)_3]$

3220–10s, b
2502vw
2268w
1460vs, b
1220w, sh
1198–5s
883m, sp, sh
805–798s
760–50inf
647–40m
546m

References 7, 9, 32, 190 (v. detailed), *358* (Xtals)

77
MANGANITE
8[MnOOH]

ca. 3550w
2800vs, vb
2660–25
2100–40vs, b
1630vw
1156–45s } b
1120–10m, sh } b
1090–80vs } b
980
605–595s, b } vb
502–00s, b } vb
449–45w, b } vb

References 3, 9, 116, 321, 391, 452 (poor), *515, 606*

78

PSILOMELANE

$2[(Ba,Mn^{II})Mn_4^{IV}O_8(OH)_2]$

800
775
715sh of plat }
585vs, b } b
535–10vs, b }
460s

References 9, 21, 116, 606

79

VARLAMOFFITE

H_2SnO_3

3400s, b
1640w
1150sh
ca. 1040vs, vb
ca. 940vb, sh

Reference 454

OTHER HYDROXIDES

		References
Barium	$Ba(OH)_2$	*4*
Chromium	$Cr(OH)_2$	*321*
Copper	$Cu(OH)_2$	*213, 312*
Euxenite		
$(Y)(Nb,Ta,Ti)_2.(O.OH)_6$		*3*
Gallium	GaOOH	*357*
Indium	$In(OH)_3$	*357*
Lanthanum	$La(OH)_3$	*4*
Limonite		
Hydrated Fe oxide		*3*
Lithium	LiOH	*4, 147, 155*

		References
Nickel	$Ni(OH)_2$	*4, 321*
Potassium	KOH	*147*
Sodium	NaOH	*147, 155*
Thorium	$Th(OH)_4$	*265*
Tungstite		
$H_2WO_4 . H_2O$		*564*
Strontium	$Sr(OH)_2$	*517*
Zirconium	$Zr(OH)_4$	*514* (OH only)
Vanadium	$VO(OH)_2$	*213*

CYANATES

No individual datum is included as very little is known to the author, but references to some compounds are given below.

		References
Silver	AgOCN	*4, 7, 32, 267*
Sodium	NaOCN	*4, 267, 441, 520* (Fulminate)
Potassium	KOCN	*7, 32, 267, 520* (Fulminate)
Ammonium	NH_4OCN	*267*
Thallium	TlOCN	*267*
Caesium	CsOCN	*267*
Rubidium	RbOCN	*267*

THIOCYANATES

As with cyanates above, no individual datum is worthy of inclusion but some references are given below.

		References
Ammonium	NH_4SCN	*7, 32, 241, 426, 461*
Barium	$Ba(SCN)_2.2H_2O$	*7, 32, 426*
Mercury	$Hg(SCN)_2$	*4, 7, 32, 426, 427*
Lead	$Pb(SCN)_2$	*4, 7, 32, 426*
Potassium	KSCN	*4, 7, 32, 94, 389, 426, 427*
Sodium	NaSCN	*7, 32, 441*

CHLORATES

80
BARIUM CHLORATE (*S*)
$Ba(ClO_3)_2 . H_2O$

3570 } s, sp doublet
3550 }
1610m, sp
989–72s
968–50vs, b
933–10s, b, sh
619–05s
509sh?
506–03s
488–83s
400w, b

References 4, 7, 32, 214, 302

81
POTASSIUM CHLORATE (*S*)
$KClO_3$

ca. 1925w?
1000–989sh
975–60vs
948–30s, sp, sh
624–19s
490–79vs with low frequency shoulder

References 4, 7, 32, 186 (Xtals), *214, 224, 302*

82
SODIUM CHLORATE (*S*)
$NaClO_3$

990vs } b triplet
980–70vs }
940–30s, sp, sh }
630–20s, sp
485–75s, sp

References 4, 7, 32, 186 (Xtals), *214, 302*

OTHER CHLORATES

The following metal chlorates are given by Rocchiccioli[302]:
Ag, Ca, Cd, Cr, Cu, Co, Mg, Nd, Ni, Pb, Sr, Zn

BROMATES

83

SODIUM BROMATE (*S*)

$NaBrO_3$

835–26sh
808–02vs
450–44s, sp
370–66s, sp (high frequency shoulder has been reported)

References 4, 7, 32, 302

84

POTASSIUM BROMATE (*S*)

$KBrO_3$

845sh
825–10sh
800–780vs, b
430–25s, sp
362–58s, sp (high frequency shoulder has been reported)

References 4, 7, 32, 302

OTHER BROMATES

The following metal bromates are given by Rocchiccioli[302]:
Ba, Ca, Co, Cu, Mg, Ni, Pb, Sr, Zn, Ag (also *Refs. 7, 32*)
Nyquist and Kagel[4] give the IR spectra of the following bromates:
Al, Ba, Cd, Cs, Li, Mg, Pb, Zn, Rb

IODATES

85

POTASSIUM IODATE (*S*)

KIO_3

805s, sh
780w, sh
760s, vb
377sh }
360m }
350sh } multiplet
330sh }
310m }

References 7, 32

86

SODIUM IODATE (*S*)

$NaIO_3$

805–00s, sh
782–77vs }
770–60vs } b
730–25sh }
385m }
375sh }
358m } multiplet
334m }

References 4, 7, 32

OTHER IODATES

The following iodates are given by Rocchiccioli[302]:
Ag, Ba, Ca, Cd, Co, Cu, Li, Mg, Mn, Ni, Pb, Sr, Zn
Nyquist and Kagel[4] give the IR spectra of the following iodates:
NH_4, Ba, Ca, Cs, Ce, Cr, Pb, Li, Ni, Rb, Ag, Sr

SIMPLE ANHYDROUS CARBONATES

(TYPE I) $A(XO_3)_n$

87

AMMONIUM CARBONATE (*S*)

$(NH)_4CO_3$

3120s
3020sh
1650s
1630sh
1395vs
1275m
1020sh
1000w
980m, sh
830m, sp
700w
660vw
645w, sp

Reference 315

88

WITHERITE

$4[BaCO_3]$

2500w
1770–1740w
1455–35s, b
1065–1040w, sp
864–50w, sp
845–34w, sh
720?w
695–85m

References 4, 7, 9, 32, 37, 56, 59, 63, 65, 110, 239, 313, 371, 400 (far-IR), *453*

89

BISMUTH CARBONATE (*S*)

$Bi_2(CO_3)_3$

1745m
1460s
1067s
858sh
846vs
820m
717w
686s

References 239, 371

90

CALCITE

$2[CaCO_3]$

2530–00
1812–1792w
1525?
1435–10vs
1087–78vw (if present at all)
881–73s
847sh
713–10w
325vs, b
230s

References 3, 4, 7, 9, 31, 32, 57, 59, 63, 65, 107, 110, 239, 313, 400 (far-IR), *422**, *518*, *608**(Xtals)

91
ARAGONITE
4[$CaCO_3$]

2530–00
1803–1770w
1493–70vs } These two intensities
1450–30sh } may be reversed
1087–80w
875–70m
860–45m, sh
715–10w } doublet
703–697w }

References 3, 9, 32, 33, 57, 63, 65, 107, 110, 143, 239, 313, 400 (far-IR), *455*

92
VATERITE
2[$CaCO_3$]

1790vw
1450–20vs, vb
1090–80w-m
880–70s, sp
860–50sh
750–40w–m
710m, sp?
330vs, b
230s, sh

References 4, 9, 57, 59

93
OTAVITE (*S*)
2[$CdCO_3$]
(Calcite Structure)

1800
1462–40vs, b
1082–75w
862–60s, sp
837–34w, sh
723m
705sh
675vw

References 4, 9, 59, 239, 371

94
SPHAEROCOBALTITE/
COBALTOCALCITE
[$CoCO_3$]

nr. 1800w
1485–80sh
1440–25s, vb
1076–70w
880–60m, sh
838–26m
747–40w
713w, sp
350m, b

References 7, 36, 59, 61, 63, 239

95

CHALYBITE/SIDERITE

2[$FeCO_3$]

2480w
2320w
1820–00w
1438–15vs, b
1087–81w, sp
873–62s, sp
740–37m
720vw
668w

References 3, 9, 33, 59, 63, 65, 110, 239, 371, 400 (far-IR)

96

CERUSSITE

4[$PbCO_3$]

1740–18w
1460–35 } vs, b doublet
1405–1390 }
1055–50w, sp
841–37m, sp
826–21vw, sh
689–73m, sp

References 4, 7, 9, 32, 37, 59, 63, 65, 110, 116, 239, 280, 371, 400 (far-IR), *429, 453*

97

LITHIUM CARBONATE (*S*)

Li_2CO_3

1510–1492s } b doublet
1450–45s }
1096–80vw, sp
865–60m, sp
741–32w
677vw?
500–480m, b
405m, b, sh
380sh

References 4, 7, 32, 36, 239, 371, 428

98

MAGNESITE/GIOBERTITE

2[$MgCO_3$]

2500vw
2300vw
1823–18vw
1575–50sh
1478–50vs, b
1086vw
900–885m, sp doublet?
855–47vw
752–48w
730–25vw
668w

References 3, 7, 9, 31, 37, 59, 63, 65, 110, 239, 371, 400 (far-IR), *533* (Xtal-reflection), *585, 608*

99

RHODOCHROSITE

$2[MnCO_3]$

nr. 2500vw
1810–1790w
1435–12vs, b (high frequency preferred)
1087–80vw
870–65s
837w, sp
728–23m
667w

References *7*, *9*, *37*, *59*, *63*, *65*, *110*, *371*, *400* (far-IR), *452* (poor), *606*

100

POTASSIUM CARBONATE (*S*)

K_2CO_3

nr. 2475vw (hydrated→doublet)
nr. 1720vw (hydrated→doublet)
1449–30vs, b
1063w, sp
875m, sp
853w
702–690w } doublet
694–80w }

References *7*, *32*, *239*, *371*, *428*

101

SODIUM CARBONATE (*S*)

Na_2CO_3

1785vw
1460vs, b
1420sh
879s, sp
848w, sp
701w, sp } doublet
695w, sp }
480w
ca. 250s, b

References *4*, *9*, *371*

102

STRONTIANITE

$4[SrCO_3]$

2520–2450vw
2300vw
1770–60w
1496–55vs, b
863–58s
847–43m, sh
711–05w, sp } doublet
703–695w, sp }
230vs, b

References *4*, *9*, *33*, *36*, *37*, *56*, *59*, *63*, *65*, *110*, *239*, *371*, *400* (far-IR), *453*

103

SMITHSONITE

$2[ZnCO_3]$

2530vw
2450vw
2280vw
1830–1790w
1440–20vs, b
1048m, b
873–70m, sp
840w, sp, sh
745–42m
670vw

References 9, 33, 37, 59, 63, 65, 110, 116, 239, 371, 533 (Xtal-reflection)

OTHER CARBONATES

Ross and Goldsmith[371] have recorded the spectra of the following metal carbonates also (additional references are given where they are known):

Cr^{III}, Cu, Bi, Be (basic), Te, Y
Cs, Cd, Co^{III} (basic), also *Ref. 4*
La, Tl, *Ref. 239*
Ni, *Refs. 37, 239*
Ag, *Refs. 4, 315*

ANHYDROUS NORMAL CARBONATES

(TYPE II) $AB(XO_3)_2$

104

DOLOMITE

$[CaMg(CO_3)_2]$

ca. 2900vw	853–47vw
2530vw	785vw
ca. 2300vw	730–25w
1818–10vw	681?
1620w	667w
1480inf	370vs, vb
1450–30vs, b	260vs, sh
1315sh	
1035 } w, b	
1010 } doublet	
896–79m (shoulder on high frequency side)	

References 4, 9, 31, 33, 59, 65, 110, 143, 239, 313, 533 (Xtal-reflection), *585*

105

HUNTITE

$2[Mg_3Ca(CO_3)_4]$

1882 vw
1547s, b }
1507s }
1480sh } b
1460sh }
1440vs }
1110w
1015–12w, b
892–89s, sp
869–67s, sp
742–40w

References 126, 143, 313

106

ANKERITE

$[Ca(FeMnMg)(CO_3)_2]$

2530w
1825w, b
1450vs, b
877s
726m, sp

References 9, 110

107

KUTNAHORITE

$2[CaMn,Mg)(CO_3)_2]$

2520vw
1818vw
1435vs, b
869s
721m

Reference 110

108

BARYTOCALCITE

$2[BaCa(CO_3)_2]$

1525sh } vb
1475vs } b doublet } vb
1420vs } b doublet } vb
1376s, sh } vb
1082w, sp
880m, sp
868m, sp
723vw, sp
693w, sp
680sh

References 10, 313

109

ALSTONITE

$2[BaCa(CO_3)_2]$

1770vw
1510sh
1470–50vs, b
1437–18sh
1087vw, sp
1065vw, sp
865–60m, sp
856sh
714sh
708–07w, sp
701sh
690–87w, sp

References 9, 313, 506

110

SHORTITE

$2[Na_2Ca_2(CO_3)_3]$

1528sh
1480sh
1454vs
1410s, sh
1089vw, sp } doublet
1070vw, sp } doublet
880sh
872sh
866m, sp
851sh
730w
718w
709sh
693w

Reference 313

111

COBALTSMITHSONITE

$2[(Zn,Co)CO_3]$

1590rev, inf
1435vs, b
1080vw, b
870m, sp
740m, sp

Reference 116

112

GASPEITE

$(Ni,Mg,Fe)_2(CO_3)_2$

1428
876
751

Reference 384 (no spectrum)

HYDRATED NORMAL CARBONATES

113
THERMONATRITE
$4[Na_2CO_3 . H_2O]$

3540m
1592vs
1429s, b
1373sh
1045w
1016w
924w
851w
768w

Reference 110

114
NESQUEHONITE
$4[MgCO_3 . 3H_2O]$

3568s, sp, sh	1518–14vs
3430vs	1470–64sh } b
3270sh	1415vs } b
3140sh	1100–1097w
2930m, sh	852
1640m, b, sh	695
1540	

References 505, 506

115
NATRON
$Na_2CO_3 . 10H_2O$

3630m, sh
3370s, b
3180s, b
1780–55vw
1640vw
1460–40vs, b
1418sp, sh
1076–65vw
937vw
905–895w
875sh
868–60m, sp
852sh
712–00sh } doublet?
695–80w } doublet?

References 4, 7, 9, 32, 371, 428, 608

116
PIRSSONITE
$8[Na_2Ca(CO_3)_2 . 2H_2O]$

3510m, vb
1493vs
1419s, b, sh
1070w
867w
711–00w, b

References 110, 313

117

GAYLUSSITE

$Na_2Ca(CO_3)_2 . 5H_2O$

3510–3450m, b
1419s, b
876–73w
718vw
691vw

References 110, 313

118

SCHROEKINGERITE

$4[NaCa_3(UO_2)(CO_3)_3(SO_4)F. 10H_2O]$

3510s, b
1558vs, b
1390vs, b
1160b, sh } b
1098s, b }
913m, b
844w, b
738w, b

Reference 110

119

ANDERSONITE

$6[Na_2Ca(UO_2)(CO_3)_3 . 6H_2O]$

3510s, b	913sh
1660inf	901s
1580vs	848w
1535sh	727w
1385vs	697w
1080w, b	
1020w, b	

Reference 110

OTHER HYDRATED CARBONATES

Nyquist and Kagel[4] have included several metal carbonates in their collection:

K, Cs, Co, Ni

CARBONATES WITH HYDROXYL OR HALOGEN

120
ZARATITE
$[Ni_3(CO_3)(OH)_4 . 4H_2O]$

3510s
1600vs, b } b
1419vs, b }
881w
835w

References 110, 315

121
HYDROZINCITE
$2[Zn_5(CO_3)_2(OH)_6]$

3300m, b
1590sh
1550sh
1515–10vs
1400–1390s, b
1340sh
1070sh } b
1050m, sp }
960–50m, b
900–890m, b
837–5s, sp
738w, sh
720sh
710w, sp
520m, sh
460 m

References 4, 9, 116, 315

122
AURICHALCITE
$4[(Zn,Cu)_5(OH)_6(CO_3)_2]$

3450sh
3405s
3300vb, sh
1560s } four peaks on a broad absorption
1505s }
1415s }
1365s }
1200w, b
1102w
1062–54m
1035m, sh
977m, b
930m, sh
835b, sh
821m
742–30m, b
705w, sh
566w, b } b triplet
505w, b }
432w, b }

References 9, 139

123
ROSASITE
$4[(Cu,Zn)_2(OH)_2CO_3]$

3490m, sp } doublet
3245m, sp }
1522s, b
1415sh
1385s
1095m
1040s, b
843s, b } b
825s }
740m } doublet
705m }

Reference 139

124
MALACHITE
$4[Cu_2CO_3(OH)_2]$

3410 } b doublet
3320 }
1510–1498vs }
1430–23sh } b doublet
1395vs }
1100–1095m, sp
900sh
885–73s, b }
830–22s, sp }
790–70m, sh } b
760–50m }
720–12sp, sh }
585sh
577–5m
527–5m
507–00m, sh
490sh
430–28w

References 3, 4, 9, 33, 110, 116, 595

125
PHOSGENITE
$4[Pb_2(Cl)_2,CO_3]$

1575sh
1518–08s, b
1350s, b
838m, sp
760m, sp
653w, sp

References 9, 116

126
BASIC BERYLLIUM
CARBONATE (*S*)
$BeCO_3 . 4Be(OH)_2 . H_2O$

1515s, sh
1430sh
1370vs
1100w
1050m
910sh
880m, b
790m
745w
670sh
650s

References 315, 371

127
BASIC CHROMIUM
CARBONATE (*S*)
$Cr_2(CO_3)_3 . Cr(OH)_3 . nH_2O$

1470s, b } doublet
1350s, b }
1075w, b
830w, b

References 315, 371

128
ARTINITE
$2[Mg_2CO_3(OH)_2 . 3H_2O]$

3610vs, sp
3230sh
3020vs, vb
1598–75s, sh ⎫
1540m, sh ⎪
1450–40vs, b ⎬ vb
1390–60b, sh ⎪
1330–20s, sh ⎭
1100w, sp

1087m ⎫
942s, vb ⎪
770s, b ⎬ vb
730–25b, sh ⎪
700–689w,sh ⎪
675s, b, sh ⎭
518m, b
444s

References 9, 381, 505

129
AZURITE
$2[Cu_3(CO_3)_2(OH)_2]$

3520m, sp
3424vw
1852vw
1509–00m, b, sh ⎫
1475–66sp, sh ⎬ b
1422–18vs ⎪
1093w, sp ⎭
955m
873sp, sh
839–35m
817s, sp, sh
770m, sp, sh
744m, sp, sh
500–490m, sh ⎫ b doublet
460–50s ⎭
408 ?

References 9, 110, 116, 595

130
HYDROCERUSSITE (*S*)
$3[Pb_3(CO_3)_2(OH)_2]$

3535w
2300–2280w
1750–35w
1420–1390vs, b
1045–40w, sp
690sp, sh
682m, sp
670sh ?

References 4, 33

131
HYDROMAGNESITE (*S*)
$2[Mg_4(CO_3)_3(OH)_2 . 3H_2O]$

3658w, sp
3525sp, sh ⎫ doublet
3463s, sp ⎭
3240sh ⎫
3010sh ⎬ b
2940m ⎭
1515–00sh
1490–75s
1122–10w
960–33vw, vb
890–85m, sp
860–45m, sp

805–792m, b
750–45w, b
722–20w, b
600s, vb
490b, sh
438s, b
420sh

References 7, 9, 33, 381, 505, 625

132
DAWSONITE
$4[NaAl(CO_3)(OH)_2]$

3285		760m, sh	b
1730w		730m, sh	b
1570s, b		690m, b	b
1403vs		650m, sh	b
1135sh	b	550s, b	
1100m	b	515b, sh	
1010b, sh	b	490s, sh	
958s	b	430v, w	
942sh	b		
865w	b		
850m, sp	b		

Reference 10

133
NORTHUPITE
$16[Na_3Mg(CO_3)_2Cl]$

1830w	
1500sh	
1475–73vs	b
1465–61s, sh	b
882–0s, sp	
713–11w	

References 10, 110, 313

134
PARISITE
$18[Ca(Ce,La)_2(CO_3)_3F_2]$

1760vw	
1460vs	vb
1435sh	vb
1095vw	doublet
1085vw	doublet
873s, sp	
748sh	
735w, sp	

References 10, 313, 399

135
STICHTITE
$3[Mg_6Cr_2CO_3(OH)_{16} . 4H_2O]$

Impure
3450s
1625m, sh
1515m, b, sh
1370s
860w, b

Reference 381 (poor)

136
SJOGRENITE
$[Mg_6Fe^{III}{}_2(CO_3)(OH)_{16} . 4H_2O]$

Impure
3450s
1615m }
1495b, sh } b
1350s }

Reference 381 (poor)

137
COALINGITE
$Mg_{10}Fe_2CO_3(OH)_{24} . 2H_2O$

Impure
ca. 3600vs
ca. 3450sh
1515b, sh }
1370s } b
870w, b

Reference 381

138
HYDROTALCITE
$3[Mg_6Al_2CO_3(OH)_{16} . 4H_2O]$

3575
1650w, b
1520sh
1400m, b
865b, sh }
650vs, vb } vb
555inf }
412vs

References 10, 381 (poor)

139
BASTNAESITE
$6[(La,Ce)(CO_3)F]$

ca. 1450vs
1100sh?
1080w
970sh
890vw
870m, sp
750w
620w
580w
545vw

References 313, 399

140

OH–BASTNAESITE

$(Ce,La,Nd)CO_3(F,OH)$

3620w
3470w
1510sh
1440vs, sp
1410m, sh
1090w, b
880w, sp
850w, sp
790w
780w
720sh
690w, b
600vw

References 313, 399

141

Th–BASTNAESITE

$ThCeCa(CO_3)F_2 . 3H_2O$

Impure
3600–3000w, vb
1520vs
1420s, sh
880m?
860s, sp
750w

Reference 399

142

TUNISITE

$NaHCa_2Al_4(CO_3)_4(OH)_{10}$

3500s, b
2650w
2330vw
1920w
1860vw
1725w
1520s, b
1155sh
1050sh
850vs, b
680s

Reference 446

ACID CARBONATES

143
NAHCOLITE (*S*)
4[$NaHCO_3$]

3080sh
2940–20s, b, sh
2500s, b
2320–00sh
2050–40vw
1925–00m
1845sh
1735sh }
1700sh }
1665–60m, b, sh } b
1630–20s }
1460–55m, sh
1410–05sh
1315–1295vs, vb
1050–48vw }
1035vw } doublet
1005–00w
840–38m
700–698s, b
662–60sh

References 7, 9, 32, 37

144
KALICINITE (*S*)
4[$KHCO_3$]

3070m, sh
2950–40s
2700sh
2620–00s
2300m, sh
1860w
1700–1690sh
1665–50vs }
1630vs } b doublet
1410–00vs
1370m, sh
1047w, sp, sh
1033w, sp, sh
1010–00m
990–80sh
833m, sp
705–02m, sp
665w, sp
475w
255m, b, sh
165vs, b

References 4, 7, 9, 32, 37

145

TESCHMACHERITE (*S*)

$8[NH_4HCO_3]$

3200–3160vs
3060vs
2850b, sh
2550m, b, sh
1890–50w, b
1715–00w, sp, sh
1690m, sh
1625–00vs
1500s
1480sh
1450?
1405–1390vs
1375–70sh
1330–25vs, b
1047–40w, sp
1030–20w, sp
1010–985m
950?
835–32s, sp
705–697s
655–49m

References 4, 7, 9, 32, 37

146

TRONA

$4[Na_3H(CO_3)_2 . 2H_2O]$

3550s
1700–1695s, vb
1472–70vs, b
1200–1190m, vb
1070–60 } w, b
1035 } w, b
851s, sp
675sh
655b, sh
600vs, vb

References 9, 110

ANHYDROUS BORATES

147
LUDWIGITE
$(Mg,Fe^{II})_2Fe^{III}BO_5$

1300–1280s,b,sh 1265–30vs 1200sh 1170sh 1050m, vb 960b, sh 880m, sh	} vb	625s, b
		560sh 520sh 480vs, vb 440b, sh } vb
800vw		
760sh		
720s, sh 670m, b	} or 710s	

References 9, 145

148
PINAKIOLITE
$(Mg,Mn^{II})_2Mn(BO_5)$

1315b, sh 1280vs, b 1210b, sh 962m, sh 885vw	} vb
808sh	
730s	
610s, b, sh 570vs, vb 470vs, vb 450sh	} vb

References 10, 145

149
KOTOITE (*S*)
$2[Mg_3B_2O_6]$

1300sh 1270s, b 1190s, b 1016–00w, b 905w, b	} vb
835sh	
810–00sh	
740inf	
720s	
640w	

References 67, 145

150
SUANITE
$4[Mg_2B_2O_5]$

1275b, sh 1180vs, vb 1075sh 1019m 976w 921w 872w 839w	} b
705s, b	
665s, b	

Reference 145

151
BEHIERITE
$4[TaBO_4]$

1070w, b
1010w, b
850m, vb
580s
510s, vb
325

Reference 640

152
SINHALITE
$4[MgAlBO_4]$

1085m, b
1020m, b
780s, b
700s, b
580m
550m
490s
400m, b
325m

Reference 640—No spectrum given

153
CAHNITE
$Ca_2[AsO_4B(OH)_4]$

945
905
840
780
545
538
435
398
275

No spectrum or band shape details given

Reference 640

154
LITHIUM TETRABORATE (*S*)
$Li_2B_4O_7$

3480	w, b doublet	1030vs	b triplet
3440		975vs	
3215w		940w, sh	
2950w, b		900vs	
1670vw		825w, sh	
1460sh	b	780w	
1420sh		720sp, sh	b quintuplet
1385sh		708sp, sh	
1350vs		690m	
1330sh		662sp, sh	
1240m, sh		650sp, sh	
1120b, sh		562w	doublet
		545w	
		510sh	
		505w	

Reference 4

HYDRATED BORATES

155
MANGANESE
TETRABORATE (*S*)
$MnB_4O_7 . 8H_2O$

ca. 3400s, b
1640w
1460ms
1380ms
1065m–s, vb (this absorption is very broad, ca. 450 cm^{-1}, and ill-resolved)

References 7, 32

156
POTASSIUM
TETRABORATE (*S*)
$K_2B_4O_7 . 5H_2O$

3560m–s, sp	1130m–s
3400s, b	1090m–s
3280m	1060sh
1640w	1002s
1465sh	930m–s
1440m	840m–s
1330m, s, b	715w
1240w	570w
1160sh	

References 4, 7, 32

157
KERNITE
$4[Na_2B_4O_7 . 4H_2O]$

3552b, sh	880sh
3340s, b	870w
1710w	832–25w
1430sh	770–68w
1375–50sb	740w, sp, sh
1290inf	725m, sh
1160 } b doublet (1160–960 } b)	710m
1145 } doublet	670m, b
1100sh	618w
1075–60vb, sh	553m
1020–10s, b	496vw
975–0s	460vw
960sh	

References 9, 145

158
BORAX
$4[Na_2B_4O_7 . 10H_2O]$

3510s, sh	1033–24sh
3365s, b	1002–996s
1650–40w	950–40s, sh
1470–65m–s	888vw?
1428–20s	830–20s
1358–40s, b	
1282–70m } doublet?	
1245m }	
1160sh	
1135–20s, sh	
1080–60m, sh	

References 4, 7, 9, 33, 67

159
TINCALCONITE (*S*)
$3[Na_2B_4O_7 \,.\, 5H_2O]$

Band	
3500sh	
3350vs	
3075b, sh	
1660w, b	
1475s	⎫
1430s	⎬ vb
1350vs	⎬
1280sh	⎭
1130s, sh	⎫
1075sh	
1025sh	
1000s, b	⎬ vb
940s, sh	
825s	
720m, sh	⎭

Reference 4

160
PINNOITE
$4[Mg(BO_2)_2 \,.\, 3H_2O]$

4610–3580s	800–780w
3440–20s	760sh
3175–70s	720w
1300–1290m	610m, sh ⎫ b
1220b, sh	587m ⎭
1165–50m, b	510m, sh
1080–70inf	490m
1035–30m	445m, vb
960m	
895m, sh	

References 9, 145

161
PRICEITE
$Ca_4B_{10}O_{19} \,.\, 7H_2O$

3590sp, sh	
3560m, sh	
3455m, b	
3100m, b	
1670w	
1430sh	
1378s	
1320–15s	
1290sh	715w
1200sh ⎫	665vw, vb
1150sh	595w
1070–66vs, b	570w, b
1020s ⎬ vb	515vw, sh
990sh	470w, b
950b, sh	
900vs, b	
790b, sh ⎭	

References 9, 145

162
ULEXITE
$2[NaCaB_5O_9 \,.\, 8H_2O]$

3600s, sh	1220–08m, b ⎫
3515b, sh	1105–1090vb, sh
3430vs	1008–00s ⎬ b
3315m, b, sh	975sh
3220m, b, sh	930–25sh ⎭
3140m, sh	868–60s
1670w ⎫ b	840sh
1638w ⎭ doublet	742–40m ⎫ b
1480sh	721–15m ⎭ doublet
1420vs	650m, vb
1360s, sh	605b, sh
1325–20s	537m, vb
	477vw
	450w

References 9, 145

163
PROBERTITE
$2[NaCaB_5O_9 . 5H_2O]$

3645m, sh		850sh
3578m, sh		840–37m
3515sh		800vw, b
3385s		757–50w
3300b, sh		725w
3100b, sh		690–70w, b
1382vs	} b	650vw
1350m, sh	} b	617w, b
1340sh	} b	558vw
1330sh	} b	517w
1216–10sh	} vb	476w, b
1187–80sh	} vb	460sh
1160–55sh	} vb	450sh
1132–30sh	} vb	
1086–80s, b	} vb	
1040s, b	} vb	
990s	} vb	
970sh	} vb	
960s, sh	} vb	
930s, sh	} vb	
920sh	} vb	
900s, sh	} vb	

References 9, 145

164
COLEMANITE
$2[Ca_2B_6O_{11} . 5H_2O]$

3603m		580vw
3320m, vb		548vw
1635vw		514vw
1365s		
1340–30sh		
1330		
1270sh		
1232–20m, sh		
1160–55m	} b doublet	
1120m	} b doublet	
1060sh	} vb	
1040s, b	} vb	
950–40s, vb	} vb	
930s, b	} vb	
892–70s, sh	} vb	
859–50sh	} vb	
815–05m,sh	} vb	
750m	} b doublet	
725m	} b doublet	

References 9, 145, 359, 533 (Xtal-reflection), *608*

165
HYDROBORACITE
$CaMgB_6O_{11} . 6H_2O$

3600s, sp	1190s
3400m, sh	1100s, b
3320m, sh	1020sh
3220s	970s
3150m, sh	900s, sh
1600vw	860sh
1390s, b	840s
1320sh	760m
1300s } b doublet	680m, b
1280s } b doublet	

Reference 9

166
MEYERHOFFERITE
$[Ca_2B_6O_{11} . 7H_2O]$

ca. 3600m, sp } b
3490vs } b
3430vs } b
3310?
3060w, b
1440m, sh
1400vs
1370–1360s, sh
1160–40vs } b quartet
1090–85vs } b quartet
1044–26vs }doublet? } b quartet
960vs } b quartet
950sh } b quartet
904–00m, b doublet
840–37m, sh
800–798m, b
730–23m, sp, sh
682–80m, vb
620–17m, b
580–70m, b
530m
500w
470w
430w

References 10, 73, 145

167
INYOITE
$2[Ca_2B_6O_{11} . 13H_2O]$

3540m } twin peaks on a broad
3450m } band centred at:
ca. 3350m, vb
1650w
1475sh
1427m, b

1333m	b	804m, sp
1230sh	b	720m, vb
1170m	b	545m, vb
1110m	b	490b, sh
1063m	b	465w, b
1006m	b	
960b, sh	b	
892m	b	

References 9, 115, 358

168
TERTSCHITE
$Ca_4B_{10}O_{19} . 2OH_2O$

3630w	1097s	5 unresolved broad peaks
3440vs	1063s	5 unresolved broad peaks
1656m, b	1020s	5 unresolved broad peaks
1444sh	973s	5 unresolved broad peaks
1375vs, b	923s	5 unresolved broad peaks
1345sh	890sh?	5 unresolved broad peaks
1332inf	844m, sh	
1268w	762m	
1229w	722w	(700–550 w, vb)
	561vw	(700–550 w, vb)
	525w	(700–550 w, vb)

References 10, 201, 277

169
KURNAKOVITE
$Mg_2B_6O_{11} . 13H_2O$

ca. 3550sh	vb	1045sh
3480–75s, sh	vb	1020s, b
3350s, b	vb	940vw
3250s, b	vb	860m, b
1708		815m, sh
1685w, b		745m, b
1465sh		660m
1415s		625b, sh
1350s		550sh
1290sh		530w, b
1260sh		470w, b
1220m		
1160w		

References 9, 115

170
LESSERITE/INDERITE
$Mg_2B_6O_{11} . 15H_2O$

3635m, sp, sh
3400 centre, s, vb band
1620m
1448m, sh
1406s
1350s
1280m
1180s, sh
1135s, sh
1050s
1000s
958sh
868s
800m, sp
745m, sp
660m, vb
585–70m, b
490w
456w } doublet
440w } doublet

References 9, 115, 145

171

LARDERELLITE

$(NH_4)_2B_{10}O_{16} \cdot 5H_2O$

3420	780s, sp
3250	745m, sp
3125	715sh
1650b, sh	696s
1450vs, b	665b, sh
1335m, sh	615sh
1285vs, b	575m
1215vs	528w
1100vs } part resolved	483sh
1062vs } part resolved	472m
1050sh } part resolved	460sh
940vs	
900sh	

Reference 10

172

HOWLITE

$Ca_2B_5SiO_9(OH)_5$

1640vw	675m, b
1460sh } b	612sh
1420m } b	595m
1365m, sh	552w, b
1329m	530 } w, b
1232m, b, sh } vb	512 } w, b
1138m, sh } vb	452m, b
1100sh } vb	
1055sh } vb	
1015s } vb	
962s } vb	
912sp, sh } vb	
902sp, sh } vb	
870sp, sh } vb	
856sp, sh } vb	
830m, sh } vb	
776sh } vb	
750sh } vb	
707vw } vb	

Reference 120

SASSOLITE

H_3BO_3

See Data Table 76

BANDYLITE

$2[CuBO_2Cl \cdot 2H_2O]$

Reference 639

RHODIZITE

$[NaKLi_4Be_3Al_4B_{10}O_{27}]$

Reference 639

BORATES WITH HYDROXYL OR HALOGEN

173
FLUOBORITE
$2[Mg_3BO_3(F,OH)_3]$

1450m, b	
1250vs, b	
880sh	
860m	
746m	
655w	b (655w, 630w)
630w	
530sh	
465s, b	b (465s, 430s, 410s)
430s, b	
410s, b	

Reference 10

174
HAMBERGITE
$8[Be_2(OH)(BO_3)]$

3595w, sp		895sh	vb (895–707)
3545w, sp		860sh	
3470m		830m, sh	
3408m		800m, sh	
3300vw		775vs	
1640vw, b		745sh	
1405m, sh		721sh	
1312vs, b		707sh	
1160m		652w	
1028vw	b doublet (1028vw, 1018vw)	603w	
1018vw		448m	
		443sh	

References 9, 533 (Xtal-reflection), *607*

175

SUSSEXITE

$(Mn,Mg)_2B_2O_5 . H_2O$

3460m	770sh
1900w	695s
1350vs, b } b	660b, sh
1270vs, b } b	
1225vs } b	
1200b, sh } b	
992m } doublet	
970m } doublet	
910sh } b	
885m } b	
820m, sp	

References 145, 596

176

ASCHARITE/CAMSELLITE/SZAIBELYITE

$Mg(BO_2)(OH)$

3552
1900
1640vw
[1485vs?] } b
[1390vs?] } b
1285–80vs } b
1270–60s, sh } b doublet } b
1235–30s, sh } b doublet } b
1020–10m, sp
985–79m, sp
930–21m, sp
850w, sp
785–75vw
710vs, sp
690–87m
630vs
570m
540m
515sh
500m
455sh
445m, b
410m

References 9, 118, 145, 202, 596

177
BORACITE
$8[Mg_6Cl_2B_{14}O_{26}]$

1422m, sh	885m, sh	600sh
1368m	860w, sp, sh	567m, sh
1295sh	850w, sp, sh	550m, sh
1205sh	840w, sp, sh	533m, sh
1188m, sh	802m, sh	480sh
1085b, sh	780sh	460m, b
998vs	710m	425sh
960m, sh	668m, b	
931m, sh	655sh	
915m, sh	620s, b	

The spectrum consists of a series of closely spaced peaks on a broad background.

References 9, 118, 608

178
Fe–BORACITE
$8[(Fe\ rich)(Fe,Mg,Mn)_6Cl_2B_{14}O_{26}]$

1405m, sh	710s
1357s	700sh
1200s } doublet	655m, b
1162s } doublet	585b, sh
1072sh	555m
1012vs, b	522m
930vs	448m, sh
900sh	426m
878sh	
852s, sh	
795m, sh	

The spectrum consists of a series of closely spaced peaks on a broad background.

Reference 118

179
ERICAITE (var. BORACITE)
$8[(Fe,Mg,Mn)_6Cl_2B_{14}O_{26}]$

1630vw	900sh	586m, sh
1405sh	878m, sh	555w
1357m, b	852m, sh	522w
1200m	794m, b, sh	448sh
1162b, sh	710w	426w
1070m, sh	662w, b	
1010s	648w, b	
982sh	625b, sh	
930m, b	610w	

Reference 9

180
STASSFURTITE
$8[(Mg,Fe)_6(Cl,OH)_2B_{14}O_{26}]$

1422–18m, sh	745sh
1370–61s	712m, sp
1210–00s, b	700sh
1186	667m, b
1164	625m
1080–68sh	585w } quartet of weak peaks
1043–40inf	568w } quartet of weak peaks
1028	552w } quartet of weak peaks
985vs	535w } quartet of weak peaks
960vs	465m, b
930vs	425vw
920–15s, sh	
890–88s, sp	
864–50sh	
840sh	
800sh	
785sh	

Series of closely spaced peaks similar to the others in the series.

References 9, 118

181

HILGARDITE

$[Ca_8(B_6O_{11})_3Cl_4 \cdot 4H_2O]$

1360s
1320s
1190sh
1150m, b
1090sh
1015s, b ⎫
980s ⎬ b
950sh ⎭
890w
870w
825w
770sh
760w
730m, b

Reference 145

182

FABIANITE

CaB_3O_5OH

3390vs, b	766sp, sh
1400vs, b	738m, sh ⎫
1333s, sp, sh	726m, sh ⎬ triplet
1146–32vs, b ⎫	716s ⎭
1060sh	677s
1040sh	667sh
1020vs	655sh
975m, sh	633m
940vs ⎬ vb	616sh
920plat	597sh
912inf	577w
865vs, sp	533m
832s, sh	506w
798sp, sh ⎭	495sh
	475w

Reference 10

COMPOUND BORATES

183
LUENEBURGITE
$Mg_3(PO_4)_2B_2O_3 . 8H_2O$

3500m, sh	1200–1185m, sh	vb
3440sh	1130–22m, sh	vb
3390vs	1090–75m, sh	vb
3250b, sh	1022–20s	vb
3130sh	980sh	vb
2980b, sh	790–85w	vb
1650m	703–00vb, sh	
1335–30sh	650–40m, sh	
1300–1285m	590–77m, b	
	539m, sh	b doublet
	523m, sh	b doublet
	465m, sh	

Reference 9

184
HARKERITE
$Ca_{48}Mg_{16}Al_3(BO_3)_{15}(CO_3)_{18}$-$(SiO_4)_{12}Cl_2(OH)_6 . 3H_2O$

1520vs		780m, sh
1510sh		770m, sp
1300b, sh		742sh
1245m, b		625vw
975s		545w
910m, sh		460m
865m, sh	doublet	
855	doublet	

Reference 393

185
HEIDORNITE
$Ca_3Na_2Cl(SO_4)_2B_5O_8(OH)_2$

3485m		722m, sp	
3430sh?		710sh	
1635vw		698m, sh	
1455sh		645m	b triplet
1420s, sh		634m	b triplet
1350vs		618m	b triplet
1238s, sh	doublet	592w	
1228s, sh	doublet	565?	
1175sh		555m, b	
1145vs		540?	
1100vs, sh		525w	
1060s, sh		497w	
1000sp?		435vw	
970s	b		
942sh	b		
872			
827s, sh			
757m, sp			
746sh			

Reference 10

186
SAGAITE
$(Ca)_{48}Mg_{16}(BO_3)_{20}(CO_3)_{16}Cl_4$-$(OH)_8 . 4H_2O$

1702vw	
1655vw	
1620vw	
1535vs	
ca. 1520sh	
1235vs	
960w	
865m	doublet
855m	doublet
780w	doublet
770w	doublet
742s, sp	
625w	
455s	

Reference 393

187
BORKARITE
$Ca_4MgH_6(BO_3)_4(CO_3)_2$

3568s	1100sh	595s
3500s	1055sh	552m
3460w, sh	1015vs	532m
3410s	960vs	512m
3225m, b	910m, sh	452s
1460m, sh	870m, sh	
1440sh	856m, sh	
1420s	830sh	
1382m, sh	776sh	
1365m, sh	750sh	
1329s	707m, sp, sh	
1232m, b, sh	690s, b	
1138m, sh	612sh	

References 9, 395

ORTHOBORATES

Duval and Lecomte[67] give the absorption positions of the following orthoborates: Mg, Ca, Ni, Mn, Sr, Ba.

The following rare earth orthoborates are to be found:

	References		*References*
$DyBO_3$	*59, 417*	$PrBO_3$	*417*
$ErBO_3$	*59, 417*	$SmBO_3$	*59, 417*
$GdBO_3$	*59, 417*	$ScBO_3$	*93, 417*
$HoBO_3$	*59, 417*	$YbBO_3$	*59, 517*
$InBO_3$	*59, 93*	YBO_3	*59*
$LaBO_3$	*59, 93*		
$LuBO_3$	*59*		
$NdBO_3$	*59*		

METABORATES

	References
$Pb(BO_2)_2 . H_2O$	*7, 67*
$Mg(BO_2)_2 . 8H_2O$	*7, 32*
$Na_3(BO_2)_3 . 4H_2O$	*7, 32, 67*
$Ba(BO_2)_2$	*67*

ANHYDROUS NITRATES

188

SODA-NITRE/NITRATINE (*S*)

[$NaNO_3$]

2400vw
2320vw
1790w–m, sp
1766w, sp
1375sh
1355–50vs, vb
1122vw, sp
839–36m–s, sp
720w

*References 4, 7, 9, 72, 159, 283, 422**, *453, 519** (*Xtals)

189

NITRE

4[KNO_3]

2400–2370w
1772–60w, sp
1390–70vs, b
826m, sp
725m

References 4 (far-IR), *7, 9, 72, 59, 159, 369, 453*

190

AMMONIA-NITRE/
NITRAMMITE (*S*)

2[NH_4NO_3]

3260vs }
3070m, sh } b
2600–2300w triplet
1760–50w, sp
1430sh
1390–80vs, b
1320–00vs, b
1045–40m, sp
833 }
825 } m, sp doublet
720–15m, sp

References 4, 7, 9, 241, 268, 453

191

NITROBARITE (*S*)

4[$Ba(NO_3)_2$]

2380w, b
1780–72w, sp
1460w, sh
1420–15m, sp
1355vs, vb
ca. 1100w, b, sh
816m, sp
730m, sp

References 4, 7, 63, 72, 315, 650

192

SILVER NITRATE (*S*)

$AgNO_3$

2350w
1770–60w, sp }
1750–40w, sp } doublet
1380–40vs, b
822–05m, sp
730–20w, doublet

References 4, 7, 72

HYDRATED NITRATES

193
GERHARDTITE (*S*)
$Cu_2NO_3(OH)_3$

ca. 3400s, b
2450w
1770w
1590m, sp
1450w, sh
1360vs, b
840w

Reference 7

194
FERRIC NITRATE (*S*)
$Fe(NO_3)_3 . 9H_2O$

ca. 3200s, b
2370w
1780w
1620m
1360vs, b
845w
810w
715w
665m, b
560w, b, sh
445m, vb
400m, sh

References 7, 32, 72

195
DARAPSKITE
$Na_3(NO_3)(SO_4) . H_2O$

3430m
1640w
1430sh
1377s
1150b, sh
1112vs, b
1060w, sh
995w
838sh } b doublet
830w, sp }

Reference 666

OTHER NITRATES

Vratny[72] has recorded the infrared spectra of the following metal nitrates:
Cu, Cs, Gd, La, Li, Mn, Hg, Nd, Ni, Pb, Pr, Rb, Sm, Sr, Th, Y, Zn, Zr, Mg
Nyquist and Kagel[4] have included the following metal nitrates:
Cs, La, Hg, Nd, Ni, Rb, Sr, Tl, Th, Y, Zn, Zr, Dy, Er, Ga, Ho, In, Sc, Tb, Tm
Brooker and Bates[650] have included the following:
Ca, Sr, Ba, Pb

ANHYDROUS SULPHATES

196
MASCAGNITE (*S*)
$4[(NH_4)_2SO_4]$

3220–10s
3135–00s } vb doublet?
3035–00 sh
1415–05vs
1187–70inf
1120–00vs, b (high frequency shoulder?)
990sp, sh
735w
622–18s

References 4, 7, 9, 32, 153, 431

197
ARCANITE (*S*)
$4[K_2SO_4]$

1310sh
1195–85sh
1150–30b, sh
1115–00vs, b
990–80w, sp, sh
728w
620–12s

References 4, 7, 9, 32, 74, 403, 435, 437, 448

198
APTHITALITE
$4[KNaSO_4]$

1190–83m, sh
1110–04vs
991–89sp, sh
621–16s
450vw

References 9, 118, 403, 431, 528

199
THENARDITE (*S*)
$8[Na_2SO_4]$

1135–30vs
1116–1095vs } Often occurs as a single broad peak with either a high or low frequency shoulder
991vw, sh
725w–m
640–35s
620–15s } b doublet

References 4, 7, 9, 32, 74, 118, 403, 431, 435

200
BARYTE
4[$BaSO_4$]

ca. 3430m, b
2350w
1650w
1470w, b
1330w
1200–1180s, sh
1127–11vs } b doublet
1089–72vs }
997–82m, sp
725w
637–30s, sh } b doublet
614–08s }

References 4, 9, 33, 63, 201, 208 (Xtals), *403, 415, 431, 435, 436, 437, 448, 585*

201
CELESTITE
4[$SrSO_4$]

1250–40sh } vb
1207–1190s } b triplet } vb
1145–31s } b triplet } vb
1096–80vs } b triplet } vb
998–93m, sp } vb
644–41s } doublet
613–10s } doublet

References 4, 9, 36, 63, 201, 403, 431, 533 (Xtal-reflection)

202
ANGLESITE
4[$PbSO_4$]

1630vw
1173–63m, sh } vb
1110–04vs } b doublet } vb
1055–47vs } b doublet } vb
969–65sp, sh } vb
629–25s, sh } b
600–592s } b

References 9, 63, 116, 403, 413, 436, 448, 533 (Xtal-reflection)

203
ANHYDRITE
4[$CaSO_4$]

1623vw
1160–49vs, b
1130–19sh } b
1100–1095sh } b
1014w, sh
677–71s
616–12m, sh
597–92s
515vw

References 9, 11, 33, 36, 63, 74, 118, 403, 431, 533 (Xtal-reflection), *608*

204
CHALCOCYANITE/
HYDROCYANITE (*S*)
$CuSO_4$

1215–05sh
1153–49s
1087–85s
1075?
962–60s
704–02m
613–06m
593–83m
500m
395m
345m

References 354 (far-IR), *415*

205
VANTHOFFITE
$Na_6Mg(SO_4)_4$

1195sh	b
1170vs	b
1142–38s, sh	b
1105–03vs	b
1005sh	b

643–40sh
627–24s
610–07sh
485w

References 9, 118

206
GLAUBERITE
$4[Na_2Ca(SO_4)_2]$

1180sh		b
1142–35vs	doublet	b
1108–01vs	doublet	b

635–33s
611–08s, sh
475w

References 9, 11, 118, 403, 431, 533 (Xtal-reflection), *608*

207
LANGBEINITE
$4[K_2Mg_2(SO_4)_3]$

1200sh		b
1175–63s	doublet	b
1155–45s	doublet	b

1106sh
1060?

663–58m	triplet
638–33m	triplet
613–08m	triplet

475w

References 9, 118, 403, 431, 608

HYDRATED SULPHATES

208
MIRABILITE (*S*)
$Na_2SO_4 . 10H_2O$

3570
3440
1635w
1190–80sh } b
1135–15vs } b
1090–80sh } b
1000–988w, sp, sh
640–35sh
623–15s–vs

References 9, 118, 448

209
LITHIUM SULPHATE (*S*)
$Li_2SO_4 . H_2O$

3470–00m, b
1625–15m
1175–70sh
1150–35sh } Absorption band may give a single peak in this region
1125–00vs, b } Absorption band may give a single peak in this region
1020–10w, sh
665
655–34m
580w, b
480m, b
440m, b
370m, b

References 4, 7, 32, 74, 403, 435, 448

210
SYNGENITE
$2[K_2Ca(SO_4)_2 . H_2O]$

3320
1685w
1195–89s, sh
1140sh
1136–25vs } doublet
1110–05vs } doublet
1002?
750m
661m } b doublet
647m } b doublet
611–03s
470vw
437w

References 118, 403, 431, 528

211
KROEHNKITE
$2[Na_2Cu(SO_4)_2 . 2H_2O]$

3580
3545
3100
1660w, vb
1185sh } vb
1154vs } vb
1112m, sh } b triplet } vb
1080s, sh } b triplet } vb
1056m, sh } b triplet } vb
995m, sp } vb
770m, b
652m
606m } vb
577m, b, sh } vb
475w
450b, sh
422vw

References 10, 431

212
LOWEITE
$Na_2Mg(SO_4)_2 . 2\frac{1}{2}H_2O$

3410	895vw
3200	644–39m, sh
3115	618–15m, b
1675w, b	
1530vw, b	
1180sh?	
1158vs	
1129vs	
1090–88s, sh	
1046sh	
996w, sh	

References 9, 118

213
BLOEDITE
$2[Na_2Mg(SO_4)_2 . 4H_2O]$

1680w, b	}	462b, sh
1165–57s, b	}	440m, b
1130–24s, b	} b	
1100–1099sp, sh	}	
998–994w, sp	}	
840–23w, b, sh	}	
720m, b, sh	}	
645sh	} b	
620–15s, vb	}	

References 9, 118, 403, 431

214
LEONITE
$4[K_2Mg(SO_4)_2 . 4H_2O]$

3410		850vw, sh
3260		720m, vb
1730vw, b		613s, vb
1640vw		455vw
1210m, sh	}	
1155sh	}	
1139vs	}	
1100sh	} b	
1084–82s, sh	}	
1005sp, sh	}	

References 9, 118, 528

215
PICROMERITE/SCHOENITE
$2[K_2Mg(SO_4)_2 . 6H_2O]$

3270	613m, vb
1730vw, b	465m, vb
1650vw	
1210m, sh	
1138s	
1103sh	
1082s, sh	
1005sp, sh	
760sh	
715b, sh	

References 118, 365, 431, 528

216

POLYHALITE

$K_2Ca_2Mg(SO_4)_4 . 2H_2O$

Various colours; absorption bands shift with changes in colour

3440
3335
3290
1670–45m
1215–04s, sh
1180sh–vs
1160–44s, b, sh–vs
1105–1095vs, b
1018sp, sh
1000–992m, sp
755m, b
660–53sh
625s
612–04m, sh
540m
510sh
485–80sh
455–50vw
445vw

References 9, 118, 431, 528, 608

217

ALUM/POTASH ALUM

$KAl(SO_4)_2 . 12H_2O$

3400
3000
1645m, vb
1230sh
1170b, sh
1105–1093vs
1075–60sh
981sh
930w, b
700vb, sh }
618sh } vb
600m }

References 9, 403, 431

218

TSCHERMIGITE/
AMMONIA ALUM

$4[(NH_4)Al(SO_4)_2 . 12H_2O]$

3400
3120
1640m, vb
1446m
1410m, sh
1250sh }
1205sh } b
1100–1094vs }
930w, b
700m, vb, sh }
618sh } vb
600m, vb }

References 9, 403, 431

219

CHROME ALUM (*S*)

$KCr(SO_4)_2 . 12H_2O$

3370–30s, b
1660–50w, b
1190sh }
1090vs } b
1025sh }
700m, vb }
615m } b
560m }

References 7, 32

220
BASSANITE
$CaSO_4 . \frac{1}{2}H_2O$

3615s
3565sh
1630–20m
1158vs
1120sh
1100–1095m, sh
1012w, sp, sh
725?
667–62s
634–30sh } doublet
605s

References 4, 9

221
KIESERITE
$4[MgSO_4 . H_2O]$

3390
3200
1665–45vw
1532vw
1200inf
1176–67b, sh
1135–26m
1047–42sp, sh
900w
775w, b
680–64vw
648–40w
628–00m, b
590m, b
460?
455?

References 4, 9, 118, 403, 415, 528

222
SZOMOLNOKITE
$4[FeSO_4 . H_2O]$

1155b, sh
1136vs, b
1093b, sh } b
1016m, sh

Reference 403

223
SELENITE
$8[CaSO_4 . 2H_2O]$

3500m
3300m
2850w
1610m
1180inf
1135vs, b

Reference 36

224
GYPSUM
$8[CaSO_4 . 2H_2O]$

3560s } b doublet
3430–20s }
2195mb
1693–80w
1630–22m, s
1150–42vs } doublet
1124–11vs }
1093w
1018–00vw, sh
720vw
673–61m
605–595m, b
470–60w, b

References 4, 7, 9, 32, 33, 175 (Xtals), *403, 413, 418, 431, 448, 510, 594, 608, 615, 616*

225
CHALCANTHITE
$2[CuSO_4 . 5H_2O]$

3250s, vb
2460m
2350sh
1650–20w
1530vw
1200–1190vs, sp } vb triplet
1150–20vs, b }
1090–70vs, b }
1020–00vs, sp
980–62m, sp
870m, b
810–790m, b
680–60m–s
630–12m } doublet
595–84m }
519w
470m, vb

References 4, 9, 32, 116, 354, 413, 431, 435, 448, 595

226
MELANTERITE
$8[FeSO_4 . 7H_2O]$

ca. 3400s, b
3330s, b
1664–45sh
1628–16m
1156–43sh
1110–1085vs, vb
990–76sp, sh
725–00m, b, sh
624–11s, b
445–40m

References 4, 7, 9, 32, 63, 413, 431, 448

227
BIEBERITE
$8[CoSO_4 . 7H_2O]$

3490
3400
3300
1686 } m, b triplet
1665–60 }
1638–28 }
1150–40sh } intensities may be reversed
1120–00vs }
1020–08m, sp, sh
992–81vw, sh
725–10b, sh } vb
672m }
635–30m–s, b }
610sh }
568s, b }
ca. 500vb, sh }
437m, b, sh }

References 4, 9, 63, 413, 431, 448

228
EPSOMITE
4[$MgSO_4 . 7H_2O$]

ca. 3400vs, vb
3270
1665–55m, b
1136–25sh
1110–1086vs, b
1040vw, sh?
998–81sp, sh
750–45b, sh
620–12m, vb
430–25m, sh

References 4, 9, 63, 74, 118, 413, 431, 448

229
GOSLARITE
4[$ZnSO_4 . 7H_2O$]

3430
3380
3300
1657sh
1645–38m, b
1200s, sh } vb
1155–36m, sh } vb
1110sh } intensities may be reversed } vb
1095–81vs, b } intensities may be reversed } vb
1022w, sh } vb
1012w, sh } vb
985–80w, sp } vb
ca. 750m, b
668–45m, sh } b
633–23sh } b
615–00s } b
435m, vb

References 9, 63, 74, 283 (Xtals), *413, 431, 448*

230
MORENOSITE (*S*)
4[$NiSO_4 . 7H_2O$]

1659m } b
1643m } b
1140sh
1110–1085vs, b (poorly resolved)
995–81sp, sh
665–41sh
626–19m
610–08sh
485sh

References 63, 74, 413, 448, 435

231
MANGANESE SULPHATE (*S*)
$MnSO_4 . 2H_2O$

3225–00m, b
1146–20vs, b
1093?
1025–18m, sh
826m, vb
660–45m
627–21m
608–00m
510m, vb

References 7, 9 (. $4H_2O$ Ilesite v. similar spectrum), *32, 413*

232

ROEMERITE

$Fe^{II}Fe^{III}(SO_4)_4 . 12H_2O$

3580
3400
1635m, b
1225s } vb
1184m, sh } vb
1145vs } vb
1070vs } vb
1002s } vb
855w
680m, b, sh } vb
618sh } vb
595s, vb } vb
494s, b, sh } vb
460sh } vb

Reference 10

233

PICKERINGITE

$MgAl_2(SO_4)_4 . 22H_2O$

3400
2960
1660m, vb
1125vs
1085m, sh
1000w, sp, sh
982w, sh
960m, sh
725b, sh } vb
638m, vb } vb

References 9, 431

234

HALOTRICHITE

$Fe^{II}Al_2(SO_4)_4 . 22H_2O$

3400
3000
1650m, b
1150sh
1121vs
1085m, sh
1068sh
1000w, sp, sh
982sh
960m, sh
725vb, sh } vb
ca. 625s, vb } vb

References 9, 431

235

DIETRICHITE

$(Zn,Fe,Mn)Al_2(SO_4)_4 . 22H_2O$

3580
3380
3000
1640m, b
1190sh
1105vs, b
957w, b
710m, vb, sh
615s, vb
600sh
528vw
520vw
459vw
420vw

Reference 10

236

COQUIMBITE

$4[Fe_2(SO_4)_3 . 9H_2O]$

3570	840vw, b
3430	670vw, b
2940	597m
1650w, b	480w
1225b, sh	455sh
1168m, sh	
1103sh	
1068s, b	
1025sh	
1014s, sh	

Reference 9

237

ALUNOGEN

$Al_2(SO_4)_3 . 18H_2O$

3575
3430
1640m, b
1118vs
1085sh
993sh
960sh
715vb, sh } vb
615s, vb }

Reference 9

ANHYDROUS SULPHATES WITH HYDROXYL OR HALOGEN

238
BROCHANTITE
$4[Cu_4SO_4(OH)_6]$

3583
3564
3400
3382
3267
3180
1630vw
1430vw
1395vw
1185sh
1128vs }
1119vs } doublet
1095–88vs, sp
988m, sp
944m, sp
ca. 900sh
873s, sp
850s, sh
780m
734m
645m, sh
630m
607sh }
602s } doublet
513m, sh
487s
473sh
421s, b

References 9, 116, 431

239
ANTLERITE
$4[Cu_3(SO_4)(OH)_4]$

3578
3484
1635vw
1157–52s, sp, sh }
1115–11vs } b triplet
1075–71s, sp, sh }
990–87w, sp }
888s, sp }
857s, sp } b doublet }
792s }
760s } b doublet } b
673sh }
641s }
618s, sh } b
612sh }
585sh }
517m }
496m } b doublet
462vw
418w

References 9, 116, 403, 431

240
LINARITE
$2[(Pb,Cu)SO_4(Pb,Cu)(OH)_2]$

3550	630m
3450	605s
1630vw	590sh
1400vw	540w
1164m, sh ⎤	497m, b
1100sh ⎥	452vw
1070sh ⎥	427w
1040s, b ⎬ b	
968sp, sh ⎥	
890vw ⎥	
824w ⎦	

References 9, 116

241
ALUNITE
$[KAl_3(OH)_6(SO_4)_2]$

3485	605m
1635vw	505sh
1170sh ⎤	475s
1086vs ⎬ vb	450b, sh
1030m, sh ⎦	
911?	
802 ⎫ m, b doublet	
780 ⎭	
700m, b	
632m, b	

[Woodhouseite (Ca)—Svanbergite (Sr) in Alunite Group, see *Ref. 609*]

References 9, 403, 431, 604

242
JAROSITE
$[KFe^{III}_3(OH)_6(SO_4)_2]$

3430
3355
3260
1650vw
1270m, sh ⎤
1190–85m, sh ⎥
1100–1087s ⎥
1028s ⎬ b
1018s, sh ⎥
900sh? ⎥
790vw ⎦
685b, sh ⎤
638m ⎥
575sh ⎥
512s ⎫ b doublet ⎬ vb
482s ⎭ ⎥
446sh ⎦

References 9, 116, 403, 632

243
ARGENTOJAROSITE
$[AgFe^{III}_3(OH)_6(SO_4)_2]$

3625	508s, b
3585	475s
3360	440m, sh
1182s, b, sh ⎤	424sh
1092vs ⎥	
1024 ⎫ s, b ⎬ b	
1012 ⎭ doublet ⎥	
675b, sh ⎦	
630m	
575sh	

Reference 10

244

CARPHOSIDERITE

$Fe^{III}_3(SO_4)_2(OH)_5 \cdot 2H_2O$

3390
3330
1215s
1085vs
1015vs
670sh
643s
596m, b, sh
507vs, b
478vs
449m, sh
422vw

Reference 10

245

PLUMBOJAROSITE

$[PbFe_6(OH)_{12}(SO_4)_4]$

3350		
1245m, sh		vb (1245–915)
1190s, sh		
1108m, sh		
1080s		
1012s	b doublet (1012, 1000)	
1000s		
915sh		
790w, b		
725vw		
642m	b doublet (642, 627)	
627m		
586m, b		
495m, sh	b (495–450)	
475s		
450sh		

References 10, 431

HYDRATED SULPHATES WITH HYDROXYL OR HALOGEN

246
LANGITE
$Cu_4(SO_4)(OH)_6 \cdot H_2O$

3580
3395
3280
1640w, b
1115vs, b
1095sh
985m
945m, sh
875s
855sh
780m
735m
(985–735: b; 780 and 735: b doublet)

631m, sh
602s
(631–602: b)
510sh
485m, b
423m, b
(510–423: b)

References 10, 431

247
SERPIERITE
$(Cu,Zn,Ca)_5(OH)_6(SO_4)_2 \cdot 3H_2O$

3400
1655w, b
1200sh
1135sh
1125s
1108s
1065sh
(1200–1065: b; 1125 and 1108: b doublet)
990w, sp
ca. 840m, vb, sh
710m, b
655sh
648m, sh
610m, b
(840–610: vb)
500sh
475w, b
445w, b, sh
422vw

Reference 10

248
KAINITE
$16[KMgCl(SO_4)_4 \cdot 3H_2O]$

3470
3355
3265
ca. 3120
1650w
1205–00sh
1183–77s, sh
1135s
(1205–1135: b)
1030–25sp, sh
910vw
ca. 800sh
760m, b
670–65sh
650–43m, sh
614–08m, b
578b, sh
528–23b, sh
472–65b, sh
415–12w

intensities vary

References 10, 118, 528

249
JOHANNITE
$4[Cu(UO_2)_2(OH)_2(SO_4)_2 \cdot 6H_2O]$

3400
1630m, b
1410w, b
1180b, sh
1120sh
1085vs, b
1010b, sh
925m, b
820m, b
(1180–820: vb)
670m, vb
545
475
440
(670–440: vb)

References 9, 116, 431

250
AMARANTITE
$Fe^{III}(OH)SO_4 . 3H_2O$

3580
3540
3390
1635m, b
1385m, sp
1225s, sh
1180m, sh
1145s, sh } vb
1065vs } b triplet
1000vs, sh
675m, sh
628b, sh } vb doublet
595s, b
507s, b } vb
480sh
445sh
422vw

Reference 10

251
HOHMANNITE
$Fe^{III}(OH)SO_4 . 3\frac{1}{2}H_2O$

3580
3545
3400
1640m, b
1229s, sh
1215sh
1130m, sh
1100sh
1085s } vb
1070s
1027sh
1008s
1000sh

670sh
655sh
625sh
600s, vb } b
530s } b
477s } doublet
420vw

Reference 10

252
FIBROFERRITE
$Fe^{III}(OH)SO_4 . 4\frac{1}{2}H_2O$

3580
3410
1652m, b
1220b, sh
1145sh
1090vs, b } vb
1005s, sh
780w
730m, sh
675m, sh } vb
650sh

580m, vb, sh
535sh } vb
473s

Reference 10

253
BOTRYOGEN
$MgFe^{III}(OH)(SO_4)_2 . 7H_2O$

3550?
3440
3300
1635m, b
1200s, sh
1142sh
1120vs } b
1105sh
1032sh
987w, sp
875w, sh

768m, b, sh
700 plat
655 } s, b
637 } triplet
617
450m, b

References 10, 431

254

COPIAPTITE

$(Fe^{II},Mg)Fe^{III}_4(OH)_2(SO_4)_6 \cdot nH_2O$

3530	750b, sh	vb
3400	675b, sh	vb
3130	602m, b	vb
1635w, b	557m, b, sh	vb
1225m, b, sh (v, b triplet)	460b, sh	vb
1140s (v, b triplet)		
1068m, sh (v, b triplet)		
998s, b, sh (v, b triplet)		

References 10, 431

COMPOUND SULPHATES

255

CALEDONITE

$2[Pb_5Cu_2(OH)_6CO_3(SO_4)_3]$

3625	955m, b, sh
3585	840vw
3410	800vw, b
1415sh } b	718vw
1402vs } b	624m } b doublet
1200sh	610m } b doublet
1125vs } b doublet	595m, sh
1105vs } b doublet	473w, b
1063s, sh	445w, b
1015w, sh	415vw
980w, sh	

Reference 10

256

CHALCOPHYLLITE

$[Cu_{18}Al_2(AsO_4)_3(SO_4)_3(OH)_{27} \cdot 33H_2O]$

3400
1640w, b
1100s, b
1030b, sh
955b, sh
815s
ca. 615m, b, sh
535vb, sh
470s, b
440b, sh

Reference 9

257

ARDEALITE

$2[Ca_2H(PO_4)(SO_4) \cdot 4H_2O]$

3575	866m } vb
3370	822w, b, sh } vb
1675m, b	725sh } vb
1145vs, b	677m } vb
1109vs, sh	630sh } vb
1013m	600s } vb
	528s } vb
	460w, sh
	424vw

Reference 10

258

BEUDANTITE

$[PbFe^{III}_3(OH)_6SO_4 \cdot AsO_4]$

3420	670sh } vb
1635w, b	625inf } vb
1160sh	600s, vb } vb
1078s	525sh } vb
1000b, sh	466vs } vb
985sh	430vb, sh } vb
900m	
800m	

Reference 10

259

SARMIENTITE

$Fe^{III}_2(AsO_4)(SO_4)(OH) . 5H_2O$

ca. 3200vs, vb
ca. 1600m
1170m, sh } vb
1100vs } b } vb
1060vs } b } vb
990sp, sh } vb
870sh
840vs, b
650s, b
570s, sh

Reference 442

260

THAUMASITE

$[Ca_6(CO_3)_2(SO_4)_2Si_2O_4(OH)_4 . 27H_2O]$

3560
3500
3470
3420
1705m } b doublet
1660m } b doublet
1400s, b
1102vs } b
1075sh } b
935vw
887w
765sh } vb
750s, b } vb
673m, sh } vb
638s, b, sh } vb
592s, b, sh } vb
555sh } vb
500m
430vw

References 10, 424

261

LEADHILLITE

$8[Pb_4(OH)_2SO_4(CO_3)_2]$

3508
1630vw
1470sh } b
1430sh } b
1410m } b
1370sh } b
1193sh } b
1115sh } b
1082vs } b
1010sh } b
841vw
682vw
623sh
612sh
602s
435m, b

References 9, 116, 431

262

HUMBERSTONITE

$K_3Na_7Mg_2(SO_4)_6(NO_3)_2 . 6H_2O$

3350s
3130m, sh
1708w
1603w, b
1400s } b doublet
1380s, sh } b doublet
1201b, sh } vb
1168vs } vb
1085vs } vb
1052s, sh } vb
986s
825w
751wv, b

Reference 667

OTHER SULPHATES

ANHYDROUS SULPHATES

	References
$Al_2(SO_4)_3$	*74*
Ag_2SO_4	*4, 411*
$BeSO_4$	*435, 448*
$CdSO_4$	*415*
Cs_2SO_4	*4, 146*
$CoSO_4$	*415*
$HgSO_4$	*4, 315*
$MgSO_4$	*415, 435*
$MnSO_4$	*415*
NH_4NaSO_4	*4, 481*
$NiSO_4$	*415*
Rb_2SO_4	*4, 146*
$SnSO_4$	*415*
$Tl_2(SO_4)_3$	*4, 411*
$ZnSO_4$	*415*
Hg_2SO_4	*4, 315*
$Fe_2(SO_4)_3$	*315*

HYDRATED SULPHATES

The spectra of the following metal sulphates are given by Nyquist and Kagel[4]:

Bi, Be, Ca, Al, Ga, In, Sb, V, Mn, Zn, Y, Ni, Cd, Hg, Ce, Pr, Nd, Sm, Eu, Gd, Dy, Ho, Er, Yb, Th, UO_2.

	References
$3CdSO_4 . 8H_2O$	*9, 68, 74, 315, 415*
$CuSO_4 . H_2O$	*7, 354, 448*
$ZnSO_4 . H_2O$	*448*
$Hf(SO_4)_2 . 4H_2O$	*374*
$Zr(SO_4)_2 . 4H_2O$	*4, 374*

MINERAL SULPHATES

Omori and Kerr[431] have obtained the spectra of a number of saline sulphate minerals, some of which have already been referred to above. The spectra are so arranged that accurate wavenumbers cannot be quoted, but the reader is referred to the work for the following minerals:

Aluminite	$Al_2(SO_4)(OH)_4 . 7H_2O$
Beaverite	$Pb(Cu,Fe,Al)_3(SO_4)_2(OH)_6$
Burkeite	$Na_6(CO_3)(SO_4)_2$
Dauberite, Zippeite	$2[(UO_2)_3(SO_4)_2(OH)_2 . 8H_2O]$
Dolerophanite	$4[Cu_2(SO_5)]$
Ettringite	$2[Ca_6Al_2(SO_4)_3(OH)_{12} . 26H_2O]$
Ferrinatrite	$NaFe^{III}(SO_4)_2 . 6H_2O$
Hanksite	$2[Na_{22}K(SO_4)_9(CO_3)_2Cl]$
Metavoltine, Mausite	$8[(K,Na,Fe^{II})_5Fe^{III}{}_3(SO_4)_6(OH)_2 . 8H_2O]$
Natroalunite	$[NaAl_3(SO_4)_2(OH)_6]$
Natrojarosite	$[NaFe^{III}{}_3(SO_4)_2(OH)_6]$
Natrochalcite	$2[NaCu_2(SO_4)_2(OH) . H_2O]$
Palmierite	$[(K,Na)_2Pb(SO_4)_2]$
Pisanite	$8[(Fe,Cu)SO_4 . 7H_2O]$
Siderotile	$2[Fe(SO_4) . 5H_2O]$
Raimondite	$Fe^{III}{}_4(SO_4)_3(OH)_6 . 4H_2O$
Sideronatrite	$Na_2Fe^{III}(SO_4)_2OH . 3H_2O$
Uranopilite	$(UO_2)_7(SO_4)(OH)_{12} . 15H_2O$
Sulphohalite	$4[Na_6ClF(SO_4)_2]$
Voltaite	$20[HK_2Fe^{II}{}_4(FeAl)_3(SO_4)_{10} . 13H_2O]$
Zincalunite	$ZnAl_3(SO_4)_2(OH)_6$

Alums and Tutton's salts may be found in *Refs. 4, 466* and Glaserite nr$[K_3Na(SO_4)_2]$ in *Ref. 528.*

SULPHITES

263

BARIUM SULPHITE (*S*)

$BaSO_3$

ca. 1260vw
1200w
1120sh
1070m
1000sh
983–51sh?
930–20vs, b
645inf
635–30s
610m, sh
513–08s } doublet
497–93s }

References 4, 7, 32, 69, 124, 302, 392

264

CALCIUM SULPHITE (*S*)

$CaSO_3 . 2H_2O$

3400m, sp
1620w
1210vw
1012w
980–70vs, b
953s?
940b, sh
655–49 vs, b
525–17m
495–87m
450–45m–s, b

References 7, 32, 302, 392

265

SODIUM SULPHITE (*S*)

Na_2SO_3

1220–14w
1140–35w
980–70vs, b
635–26s
496–93s

References 4, 7, 32, 392

266

POTASSIUM SULPHITE (*S*)

$K_2SO_3 . 2H_2O$

ca. 3400w
1180–75sh
1129–20sh
1100vs, b
990inf
965–50vs, vb
620vs, sp
480–78m, sp

References 7, 32, 392

OTHER SULPHITES

Rocchiccioli[69] quotes the absorption positions of several sulphites: Na, NaK, Sr, Ag, Cd, Pb, UO_2; some others are given by Nyquist and Kagel[4].

SPINELS

267
CHRYSOBERYL
$4[BeAl_2O_4]$

1161s
1102sh
1052sh
1008sh
963sh
796m
779sh
754sh
641m, b
606sh
582m, sh
571m
550sh
535m
515m
473w, b

References 3, 9, 172

268
SPINEL
$8[MgAl_2O_4]$

1080sh
1020m
765–40b, sh
690–85s, b
583–70sh
538–10s, vb
309–07m

References 9, 433, 495, 646

269
HERCYNITE (*S*)
$8[FeAl_2O_4]$

672vs }
562s } b
510s }

Reference 495

270
GAHNITE
$8[ZnAl_2O_4]$

750b, sh
690sh
670–66s, b
650sh
590inf
562–57s, b
510–05s b

References 9, 495, 533 (Xtal-reflection)

271
GALAXITE (*S*)
$8[MnAl_2O_4]$

650vs }
530m }
485s, b } vb
465s, b }

Reference 495

272
MAGNESIOFERRITE (*S*)
$8[MgFe_2O_4]$

582vs, b
433s, b

References 207, 495

273
MAGNETITE
8[Fe_3O_4]

575–00s, b
410–380s, vb (lower frequency preferred)
268
178

References 3, 4 9, 174, 407, 416, 591

274
ULVOSPINEL
$TiFe_2O_4$

ca. 550s, vb

Reference 495

275
FRANKLINITE
8[$ZnFe_2O_4$]

670sh
565–50s, vb
435m, b
393–75s, b
330–25m, b

References 3, 9, 207, 495, 591

276
TREVORITE
8[$NiFe_2O_4$]

602vs } b
ca. 530sh } b
410s, b

References 207, 495

277
CHROMITE
8[$FeCr_2O_4$]

700sh
635–18m, b
524–00m, b
435b, sh

References 9, 495, 533 (Xtal-reflection), *590*

278
FERROPICOTITE
8[(Fe,Mg)(Al,Fe)$_2$O$_4$]

660vs, b
600vs, b
510w, sh
450vs, b

References 533 (Xtal-reflection)

OTHER SPINELS

	References
$CdAl_2O_4$	*495*
$CoAl_2O_4$	*495*
$CuCr_2O_4$	*4, 495*
$CoCr_2O_4$	*495*
$MgCr_2O_4$	*463, 495*
$NiCr_2O_4$	*495*
$ZnCr_2O_4$	*169, 495*
$CuMn_2O_4$	*229*
$NiMn_2O_4$	*229*
$CoFe_2O_4$	*207, 495, 591*
$CuFe_2O_4$	*495*
$NiFe_2O_4$	*207, 495, 591*
$FeCo_2O_4$	*495*
Picotite	
$8[(Fe,Mg)(Al,Cr)_2O_4]$	*590*

SELENATES

279
SODIUM SELENATE (*S*)
$Na_2SeO_4 . 10H_2O$

3720w?	846vs	b
3500m, sp	800s	b
3200m	790sh	b
2350m, b	740sh	b
1720w, b	628s	doublet
1630w	613s	doublet
1390s	440vs	doublet
1240w, b	417vs	doublet
1125w	395sh	
1100w doublet	365w	
880s		

References 4, 7, 32, 148, 309, 437

280
POTASSIUM SELENATE (*S*)
K_2SeO_4

3300w		840sp, sh
2400–2300w multiplet		820m
1755w, sp		810sh
1370s, b		720m
1140w	triplet	420s, sp
1110w	triplet	
1080w	triplet	
890sh		
875w		
855w		

References 4, 7, 148

281
CALCIUM SELENATE (*S*)
$CaSeO_4$

1623s	
1148w	b
1125w	b
1100w	b
922b, sh	b
902vs	b
874b, sh	b
851m, sh	b
690m, b	

References 63, 309

282
BARIUM SELENATE (*S*)
$BaSeO_4$

1691w	
1144vw	
1111w	doublet
1086w	doublet
937m	
873vs, sh	doublet
852vs	doublet
839sh	
670w, b	

References 63, 148, 435

OTHER SELENATES

Duval and Lecomte[148] have included the following metal selenates:

	Zn, Be,	Mo,	Al, Co,	NH_4,	Pb
Other references		*4,*		*4, 7, 32*	*63*

Also	*References*	*Also*	*References*
Ag	*4, 309*	Cu	*4, 7*
Cd	*309*	Fe	*4*
Sr	*63*	Ni	*4*

SELENITES

283
SODIUM SELENITE (*S*)
Na_2SeO_3

800–790s, sh
740–35s, b
470–55s } b doublet
390s }

References 4, 7, 309

284
BARIUM SELENITE (*S*)
$BaSeO_3$

802s, sp, sh
765sh
737vs
616vw
590w
543w
470sh
463–53vs
418–09s
376s, sh

References 4, 302, 309

285
CHALCOMEMITE (*S*)
$4[CuSeO_3 . 2H_2O]$

3300m, b
3150m
2850m
ca. 2300w
1660w
1560w
1020w, sh
920w
780w, sh
720s, b
570?

References 7, 309

OTHER SELENITES

	References
Zr	*302*
Zn	*4, 309*
Cd	*309*
K	*4*

TANTALATES, NIOBATES AND TITANATES

286
TANTALITE
$4[(Fe,Mn)(Ta,Nb)_2O_6]$

860sh } unresolved
830sh } broad
690–499 vs, vb } band

References 116, 603, 624

287
COLUMBITE
$4[(Fe,Mn)(Nb,Ta)_2O_6]$

880sh } unresolved
830–10b, sh } broad
700–560vs, vb } band

References 9, 116, 603, 624

288
POTASSIUM NIOBATE (*S*)
$KNbO_3$

660vs
550b, sh
375m

References 4, 231

289
SODIUM NIOBATE (*S*)
$NaNbO_3$

675vs, b
510sh
375w, b

Reference 231

290
ILMENITE
$2[FeTiO_3]$

ca. 1000w, b
700–690b, sh
545–32vs, b
461–44s, b, sh
325–295s?

References 3, 7, 9, 21, 407, 465, 612

291
PEROVSKITE
$8[CaTiO_3]$

1450w, vb
875vw, sp
725–00b, sh
570–40vs, vb
450–30s, sh
380–60s, b

References 3, 4, 9, 231, 624

292
BARIUM TITANATE (*S*)
$BaTiO_3$

Tetragonal	Hexagonal
545	550s, b
400	365m

Distinct differences between spectra obtained from discs and single crystals

*References 4, 71, 180**, *231** (*Xtals)

OTHER TITANATES

Last[231] also gives the following titanates:
Sr, Cd, Pb, Mg, Zn

Nyquist and Kagel[4] also quote the following titanates:
Sr, Pb, Zn, Li, Bi, Co, Ni, Cu, Ce, Eu

Pilipenko *et al.*[624] have also included:

Ilmenorutile	$2[TiO_2]$
Priorite	$n(Yt,Er)(Nb,Ti)_2O_6$
Aeschynite	$(Ce,Th)(Nb,Ti)_2(O,OH)_6$
Euxeanite } Fersmite }	$(Ca,Ce)(Nb,Ti)_2(OF)_6$
Samarskite	$n(Yt,U,Fe,Th)(Nb,Ta)_2O_6$
Pyrochlore	$8[(Ca,Na,Ce)_2(Nb,Ti,Ta)_2(O,OH,F)_7]$

ARSENATES

293

BERZELIITE

$8[(Ca,Na)_3(Mg,Mn)_2(AsO_4)_3]$

1640vw
1075w, b
1028vw
850s } vb doublet
802s }
490sh
440s, b
410m, sh

Reference 10

294

CARYINITE

$(Ca,Pb,Na)_5(Mn,Mg)_4(AsO_4)_5$

1050w, b
832s, b
427sh
418s, b

Reference 10

295

PHARMAKOLITE

$4[CaMgH(AsO)_4 . 2H_2O]$

3380
1660 } w, b
1630 }
1265vw
900m, sh
865s
817m, sh
725w
ca. 600vw, vb
460b, sh } b
433m }

References 9, 615, 616

296

ROSELITE

$2[\alpha\text{-}Ca_2(Co,Mg)(AsO_4)_2 . 2H_2O]$

2900
1550w, b
1100sh }
985s, sh }
920sh } vb
870sp, sh }
850m, sh }
805vs
535sh }
453sh } b
435vs }

Reference 10

297
ERYTHRITE
$2[Co_3(AsO_4)_2 \cdot 8H_2O]$

3445s, sh	
3180vs, b	
1670vb, sh	
1580w, vb	
950w, sh	b
825–18sh	b
783vs	b
555w, b	
490m, b, sh	
452m, b	
428m, b	

References 3, 4, 9, 116

298
CABRERITE
$2[(Ni,Mg)_3AsO_4 \cdot 8H_2O]$

3445		575m, vb	vb
3150		510m	vb
3030		464m	vb
1685w, b		427m	vb
1595m, b			
925m, sh	b		
895m, sh	b		
843vs	b doublet; b		
792vs	b doublet; b		
720s	b		

References 9, 116

299
ANNABERGITE
$2[Ni_3(AsO_4)_2 \cdot 8H_2O]$

3430	
3160	
1630w, b, sh	
1600w, b	
1025m, b	vb
935b, sh	vb
870sh	vb
832vs	b doublet; vb
795vs	b doublet; vb
605m, vb	vb
510m	vb
460m	vb
427m	vb

References 9, 116

300
KOETTIGITE
$2[Zn_3(AsO_4)_2 \cdot 8H_2O]$

3440		565vb, sh	vb
3185		498s	vb
3045		455s	vb
1700sh	b	425s	vb
1645sh	b		
1590m	b		
890m, sh	vb		
868sh	vb		
828sh	vb		
790vs	vb		
720b, sh	vb		

Reference 10

301
SYMPLESITE
$2[Fe_3(AsO_4)_2 . 8H_2O]$

3385
1635w
1082vw
1035vw }doublet ?
1015vw }
850m, sh }
780s }b
736m, sh }
525b, sh
465b, sh
445m, b

Reference 9

302
SCORODITE
$8[Fe(AsO_4) . 2H_2O]$

3520s, sp
3020s, vb
1595w
890sh
840–25vs
730m, sh
585–58m, b
498–85s }
472–65s }triplet
440–30s, b }
390m, sh
340sh
320m
300m, sh
270m
250m, sh
240sh ?
210vw

References 4, 9, 116, 440

303
SYNADELPHITE
$10[Mn_4(OH)_5(AsO_4)]$

3580
3400
1600vw, b
1435w, b, CO_3 ?
1080b, sh
950m
870sh }
825sh }
810s }b
780m, sh }
720vw
642inf }
620s, b }b
580b, sh }
530b, sh
445s, b
410b, sh

Reference 10

304
ALLACTITE
$2[Mn_7(AsO_4)_2(OH)_8]$

3570
3480
3440
3290
1440w, b
1080b, sh }
1012w }
950sh ? }
871s }vb
860sh }
830b, sh }
760m }
660b, sh }b
637m, b }
473sh ?
440s, b
431sh
415sh

Reference 10

305
ATELESITE
$Bi_8(O)_5(OH)_5(AsO_4)_3$?

1085w
1020vw
975sh
950w
870sh }
844sh } b
822sh }
785s }
575m, b
508w
470sh
420m, b

Reference 10

306
SARKINITE
$8[Mn_2(OH)AsO_4]$

3520
1430CO_3?
1075w?
1020w?
870sh }
850m, sh }
825s } b
810sh }
725sh }
484m, b, sh }
435m, vb } b
415sh }

Reference 10

307
OLIVENITE
$4[Cu_2(OH)AsO_4]$

1640vw, b
1082vw
1037vw
938b, sh
860s } partially
828s } resolved
790s } triplet
638vw
600sh
545s, b
493w, vb
452w, vb

References 10, 116, 227 (Adamite–Olivenite series), *597*

308
ADAMITE
$4[Zn_2(OH)AsO_4]$

3580
3555
3510
885–80m, sh
826inf
812–05vs
794sh
740–30sh
535s
475m, b, sh
410?

References 9, 227 (Adamite–Olivenite series), *597*

309

MIMETITE

$2[Pb_5(AsO_4)_3Cl]$

1630vw, b	560–50vw, b
1010–999vw	419–18vs
980w	385vs
960–52w	319m, b
916vw	217w, b
868sh	134s, b
814–10vs } b doublet	
804sh } b doublet	
790–82vs } b doublet	

References 9, 116, 257, 432, 509, 597, 598, 600

310

SVABITE

$2[Ca_5(AsO_4)_3(F,OH)]$

(Impure)

(1448)CO_3
(872)CO_3
858
840s
(725)
455s, b
415b, sh

Reference 10

311

LIROCONITE

$4[Cu_2Al(OH)_4AsO_4 . 4H_2O]$

3515	635b, sh } b
3330	580m, sh } b
1640w	530s, b } b
1182w	415m
1100s, sh	
1026s	
890sh } b	
876sh } b	
828sh } b	
801vs } b	

Reference 10

312

TYROLITE

$Ca_2Cu_9(OH)_{10}(AsO_4)_4 . 10H_2O$

3490	880sh
3380	860–54m, sh
3060?	812s
1630w, b	665vw } b
1400vw	640vw } b
1200w, sh	620vw } b
1138sh	550b, sh
1126sh	475m, vb
1100w	430w, b
1035sh	
950w, b, sh	

References 9, 116

313
BAYLDONITE
$(PbCu)_7(OH)_2(AsO_4)_4 . H_2O$

1400w, b
1180vw
1090w, b
1052m, sh
890m, sh } b
805s } b
684w
475s, vb

Reference 10

314
EUCHROITE
$4[Cu_2(OH)(AsO_4) . 3H_2O]$

3535m } b doublet
3460m } b doublet
1650w
1610b, sh
1035m } b
990m, sh } b
870sh } b
830s } b
770vs, sh } b
698w, b } b
534m, sh } b
496s, b } b
415m } b

Reference 10

315
TROEGERITE
$H_2(UO_2)_2(AsO_4)_2 . 8H_2O$

3410
1645m
1034m } b
908m, sh } b
825vs } b
710vw
610vw
550w
470w

References 9, 116

316
ZEUNERITE
$Cu(UO_2)(AsO_4)_2 . 16H_2O$

3320
1640w, b
1028s, b
1005sh } b
932m, sh } b
885m, sh } b
815v, s } b
690w, vb
555w
470w

References 9, 116

317
PHARMACOSIDERITE
$[KFe^{III}_4(OH)_4(AsO_4)_3 . 6\text{–}7H_2O]$

3420
1630m, b
1040b, sh
943sh } b
900m, sh } b
825s } b

Reference 9

OTHER ARSENATES

	References		References
$Ca_3(AsO_4)_2$	*7, 32, 510*	$GaAsO_4 . 2H_2O$	*331, 440*
$(NH_4)_2AsO_4$	*315*	$Cu_3(AsO_4)_2 . 4H_2O$	*4*
$AlAsO_4$	*331*	$Zn_3(AsO_4)_2 . 8H_2O$	*4*
$GaAsO_4$	*331*	$Hg_3(AsO_4)_2$	*4*
$BAsO_4$	*331*	$SbAsO_4$	*4*
KH_2AsO_4	*4, 7, 32*	$(NH_4)_2HAsO_4$	*4*
Na_2HAsO_4	*4, 7, 32*	$PbHAsO_4$	*7, 32*

VANADATES

318
CARNOTITE
$2[(K,Na,Ca,Cu,Pb)_2(UO_2)_2(VO_4)_2 \cdot 3H_2O]$

3440
1635w
1410vw, b
1100vb, sh
1030m, sh
988m
900m
745m, b } b
635sh } b
580s, vb } b
480s, vb } b

References 9, 116, 630

319
SILICOCARNOTITE (*S*)
$4[Ca_5SiP_2O_{12}]$

1065vs } vb
1030b, sh } vb
950s, b, sh } vb
890sh } vb

Plotted point by point

Reference 177

320
TYUYAMUNITE
$4[Ca(UO_2)_2(VO_4)_2 \cdot 10H_2O]$

3440
1632m
1410vw
1075sh } b
1020sh } b
985s } b
885m
805sh } vb
746m, sh } vb
640b, sh } vb
580m, vb, sh } vb
472s, b } vb

References 9, 116

321
PUCHERITE
$4[BiVO_4]$

1635w, b
1170b, sh
1090s, b
825s, b
808sh
610s, vb
515w, vb, sh?
475m, b, sh
423w

Reference 10

322

BRACKEBUSCHITE

$2[Pb_2(Mn,Fe)(VO_4)_2 \, . \, 2H_2O]$

ca. 1050vw, b }
895m, sh }
860s, sh } b
740s, b }
730sh }
678s, b, sh }
550vw, b
495w, sh } b
475w, b }

Reference 10

323

CORVUSITE

$(VO)V_6O_{16} \, . \, nH_2O$?

1650–40w
1085sh?
1032vs
530m, b, sh
470s
440sh

References 9, 116

324

DESCLOIZITE

$4[Pb(Zn,Cu)(OH)(VO_4)]$

1640vw
1020vw
922–15m, sh }
860–54s, sp } vb
750–42vs, b }
600b, sh }
506vw?
453w, b

References 9, 116, 408, 618

325

MOTTRAMITE

$4[Pb(CuZn)VO_4 \, . \, OH]$

994m
936w
839s } doublet } b
814s } }
738sh }
704vs }
486m }

References 408, 630

326

VANADINITE

$2[Pb_5Cl(VO_4)_3]$

960–52w	536vw
870vw	419m
840–30sh	383m, sh
810–01vs } b	374m
745–37vs } doublet	319m
650sh	222m
558–0w, sh	150s, b
	132s, b

References 9, 116, 257, 408, 509, 597, 598, 600, 630

327

VOLBORTHITE

$Cu_3(VO_4)_2 \, . \, 3H_2O$

3410	678m, b
3250	605b, sh
3030	550m, b
1635w, b	470b, sh
1132m, sh } b	427s, b } doublet?
1046sh }	414s, b }
1020vs }	
1000sh }	
920b, sh }	

Reference 9

OTHER VANADATES

Ortho

Nyquist and Kagel[4] include the following vanadates in their collection: Na, Ca, Pb, Fe, Ag, $(Ca,Ni)_5OH(VO_4)_3$, $(Ca,Cu)_5OH(VO_4)_3$

Meta	*References*	*Minerals*		*Reference*
NH_4	*4, 7, 32*	Cuprodescloizite	$4[Pb(Cu,Zn)VO_4 . OH]$	*630*
Na	*7, 32*	Metarossite	$CaV_2O_6 . 2H_2O$	*630*
Cr	*568*	Metahewettite	$CaV_6O_{16} . 9H_2O$	*630*
Fe	*568*	Pascoite	$Ca_2V_6O_{17} . 11H_2O$	*630*
Co	*568*	Chevetite		*630*
Ni	*568*			

PHOSPHATES

328
MONETITE (*S*)
$CaHPO_4$

ca. 2800m
ca. 1620m, vb
1390m } b
1350m }
1170sh
1130vs
1068vs
ca. 1000m, b, sh
900m, b

References 47, 152, 153

329
POTASSIUM
ORTHOPHOSPHATE (*S*)
(dibasic)
K_2HPO_4

3400sh
ca. 2900s, vb
2500m, b
ca. 1950m, vb
1328m, sp
1300?
1110vs, b
996sh } b
983sh }
970–40s, b
850–30s, b, doublet?
625w, b
560sh
545m
520sh

References 4, 7, 32, 47, 76, 418

330
TRIPHYLITE
$4[Li(Fe^{II},Mn)PO_4]$

1138s, sh
1095sh
1045vs
980m, b, sh
640m
578m
550m
500b, sh
468m, b

Reference 9

331
LITHIOPHILITE
$4[Li(Mn,Fe)PO_4]$

1138–25s, sh
1094sh
1075–52s, b
1000–990m, b, sh
825w, sh
638m
578s, sh } b doublet
551m }
490b sh
463m, b

References 9, 375

332
NATROPHILITE
$4[Na,Mn(PO_4)]$

ca. 1120sh }
ca. 1085sh }
ca. 1050vs } vb
ca. 970b, sh }
ca. 940sh }

Reference 375

333
SICKLERITE
$(Li,Mn^{II},Fe^{III})PO_4$

ca. 1085s, b } vb
ca. 1025b, sh }

Reference 375

334
ALLUAUDITE
$(Na,Fe^{III},Mn^{II})PO_4$

ca. 1100s, b
ca. 1030s, b } vb

Reference 375

335
HETEROSITE
$4[(Fe^{III},Mn^{III})PO_4]$

1650w, vb
1100sh
1090vs
1035vs, b
970s, sh } vb
685b, sh
655m
605sh
580m, sh
537s, b
410m, b

Reference 10

336
PURPURITE
$(Mn^{III},Fe^{III})PO_4$

1640vw, b
1068s
1035m, sh
990m, b, sh } b
690sh
655sh
575sh
535m, b
420m, b } vb

Reference 10

337
BERYLLONITE
$12[NaBePO_4]$

1150m, sh
1115sh
1075m, sh
1058vs, sp
1022sh } vb
790sh
773s
740s, sh
725sh
690s, b, sh } b
639w
588m, sh
576s
563sh
550sh
535m, sh
510m, sh
485sh
465m, sh
452sh } vb

Reference 10

338
ARROJADITE
$12[Na_2(Fe,Mn)_5(PO_4)_4]$(?)

ca. 1100s } vb
ca. 1000b, sh } vb

Reference 375

339
WHITLOCKITE
$7[Ca_3(PO_4)_2]$

ca. 3200m, b
1630w
1093–90sh
1070inf
1045–30vs
962sp, sh
860w, sp
725w–m
635–30m, sh } b triplet
605s } b triplet
575–65s } b triplet
475w
455m

References 4, 7, 32, 47, 150, 315

340
GRAFTONITE
$4[(Fe^{II},Mn^{II},Ca)_3(PO_4)_2]$

1040m, b } vb
1000sh } vb
630b, sh } b
595b, sh } b
550s, b } b
500s, vb } b

References 3, 10

341
XENOTIME
$4[YPO_4]$

900s, vb
570m, b, sh
500 ?b, sh
440s, vb

References 3, 9, 21, 533 (Xtal-reflection), *612*

342
MONAZITE
$4[(Ce,La,Dy)PO_4]$

1100–1050vs } b } vb
1080vs } b } vb
970–55sh } vb
905vw, sh } vb
625–20m, sh } b
585–65sh } b
545–40sh } b
475w } b

References 3, 9, 21, 394, 612

343
BERLINITE (*S*)
$3[AlPO_4]$

1100–1035vs, vb
750w, sp
ca. 700s
475s, b
450sh

References 4, 150, 331, 376

344
BRUSHITE (*S*)
$4[CaHPO_4 . 2H_2O]$

3450m
ca. 3200b
2300b
1700b, sh
ca. 1130vs }
1050vs } vb
990vs }
880s, vb
585sh
568m
530w, b
408w

References 7, 32, 33, 615, 616

345
NEWBERYITE (*S*)
$MgHPO_4 . 3H_2O$

3550s, sh	875–70m, b
3400sh	640sh
3275vs	615sh
2500sh	600m
ca. 2300w, b	590 ?w
1680sh	580m
1640m	510w
1240m	433w
1160s	415w
1050vs } doublet	395w
1020s } doublet	

References 7, 32

346
STRUVITE
$2[NH_4Mg(PO_4) . 6H_2O]$

3280
3010
1680 } w, b
1635 } w, b
1472sh
1440m
1410sh
1012vs, b
905b, sh
768m
700b, sh
576m–s
465w, b

References 4, 9, 47

347
DICKINSONITE
$[H_2Na_6(Mn^{II},Fe^{II})_{14}(PO_4)_{12} . H_2O]$?

3630
1080sh }
1030s, b } b
990inf }
925b, sh }
830vw
755w
537s, b }
480s, b } vb
420m, b, sh }

References 10, 375

348
COLLINSITE
[$Ca_2(Mg,Fe)(PO_4)_2 . 2H_2O$]

3030
1650w, b
1530?
1112s, sh }
1000vs, b }
945m, sh } vb
920sh }
770s, sh }
577s } b
562sh }

Reference 10

349
HOPEITE (*S*)
4[$Zn_3(PO_4)_2 . 4H_2O$]

3530m, sh	722w
ca. 3150vs, vb	630s
1640m	575m, sh
1145sh } b } b triplet	
1125sh } b }	
1105vs } b }	
1075sh } b }	
1020vs }	
940vs }	

Reference 4

350
VIVIANITE
2[$Fe_3(PO_4)_2 . 8H_2O$]

3480–00m, sh
3200?
3065s, b
1635–1595m, b
1040vs, b } doublet?
990–70m, sh }
890w, sh
872w, sh
794?
590m, b
560b, sh
475vw

References 3, 4, 9, 21, 104, 375, 458 (thermal studies), *533* (Xtal-reflection)

351
BARRANDITE
8[$(Fe,Al)PO_4 . 2H_2O$]

3450s, sp
3080vs
1640w
1190m, sh
1065vs, b
945w
ca. 800w, b

Reference 376

352
CUPRIC PHOSPHATE (*S*)
$Cu_3(PO_4)_2 . 3H_2O$

3400w, b
2300w
ca. 1600sh
ca. 1520w, b
1300m, b
1150m, sh } b
1100sh } b
1070vs, vb } b
1015vs } b
995sh } b
960s } b
930s } b
860m, b
645m, sh } doublet
620m } doublet
562m
510w, b
430w
400w
350sh
325w

References 4, 7, 32

353
NICKELOUS PHOSPHATE (*S*)
$Ni_3(PO_4)_2 . 7H_2O$

3400m, sp
ca. 3100vs, b
2320w
ca. 1590w, b
1060sh
1005vs, b
950sh
880m, b
745w, vb
830m, sh
575m, b

References 4, 7, 32

354
MAGNESIUM PHOSPHATE (*S*)
$Mg_3(PO_4)_2 . 4H_2O$

3400w, sh
3250m, b
2350w, b
ca. 1625m, b
1230w
1150sh
1130s, sh
1080sh
1050sh
1010vs, vb
980inf
930m, sh
880w, b, sh
780w
605m, b
570m, sh
490w
445sh
430m
350m

References 4, 7, 32

355
MANGANESE PHOSPHATE (*S*)
$Mn_3(PO_4)_2 . 7H_2O$

3400m, sp } triplet
3330m, sp } triplet
ca. 3150m } triplet
2450w, b
2300w
1295w, b
1250w, b
1145m, sh
1090sh
1070sh
1040sh
1025vs, vb
975m, sh
935sh
905w, sh
600sh
580m
530m
475sh
450sh
375w
335m, sh
315m

References 7, 32

356

STRENGITE

$8[FePO_4 . 2H_2O]$

3565	650w, b ⎫
3130	586m ⎬ b
1610vw, b	545m, b ⎭
1125m, sh ⎫	520sh
1024vs ⎬ b	485vw
990sh ⎭	475vw
870vw	430w
760vw, b	

References 10, 47, 315, 376

357

METASTRENGITE/
PHOSPHOSIDERITE

$4[FePO_4 . 2H_2O]$

3330m, sh
3060m
1665m
1150sh ⎫
1080sh ⎪
1040vs ⎬ vb
1010sh ⎭
765w, b

Reference 376

358

VARISCITE

$8[AlPO_4 . 2H_2O]$

3588–70sp
3110–3030s, b
2945–40sh
1640w
1585m
1175–60m, sh
1100–1075vs ⎫
1075–50vs ⎬ b doublet?
940–38m
885–70m ⎫
800m, b ⎪
710–05m ⎪
646m ⎪
612m ⎬ vb
577m ⎪
515m ⎪
450s ⎪
420sh ⎭

References 9, 376

359

WEINSCHENKITE/CHURCHITE

$4[(Y,Er)(PO_4) . 2H_2O]$

3570
3400
3140
1716w
1640m
1070vs, b ⎫
1020vs, b ⎬ doublet
978sh?
753m
650s
533s, b

References 10, 615, 616

360
MORINITE
$Na_2Ca_4Al_4(PO_4)_4O_2F_6 . 5H_2O$

(approx.)
1135sh }
1120sh }
1100sh } vb
1050s }
1010m, sh }
895m

Reference 375

361
CORNETITE
$8[Cu_3(OH)_3PO_4]$

3285m } b	647m, sh
3200m } doublet	623m
(1455)CO_3 ?	582m
1395w	558m, sh
1150sh }	510m, sh
1070sh }	464s, b
1015plat } vb	415m, sh
1000vs }	
960m, sh }	
850m, sh }	

Reference 10

362
PSEUDOMALACHITE
$2[Cu_5(OH)_4(PO_4)_2 . H_2O]$

3582	762w
3435	640sh
3390	615m, b
1638m, b	555vw
1500vw	530vw
1390vw	482vw
1325vw	450vw
1180b, sh	422vw?
1096m	
1025m, b	
953m, sh	
890vw	
810w	

References 9, 116

363
HERDERITE
$4[CaBe(OH,F)PO_4]$

3605	654vw?
3570	638w
3430	621w, sh
1147–36vs } b	576–71m, sh
1110–05vs } triplet	557–50s
1081–79s, sh }	483m, sh
1029–25vs, b	467m
1014–10sh	422w
905–899w	
834–30b, sh	
790–85s	
732s, sh	
705sh	

References 10, 172

364

AMBLYGONITE

$2[(Li,Na)AlPO_4(F,OH)]$

1188m, sh
1105vs, b
1028m, sh
815m
630b, sh
600s, b
540m, sh
485s, b
420s

References 9, 36

365

MONTEBRASITE

$2[LiAl(OH)PO_4]$

3350
1190m, sh } b
1100–1085vs, b } b
1030–10s, sh } b
832–20m, b
655m, b, sh } vb
600sh } vb
575w, sh } vb
545vs, b } doublet } vb
490vs, b } doublet } vb

References 10, 375

366

CRANDALLITE

$[CaAl_3(OH)_5(PO_4)_2 . H_2O]$

3580
3320
3140
1330vw } vb
1225sh } vb
1175sh } vb
1140sh } vb
1120m } vb
1080sh } vb
1055sh } vb
1040m } vb
871w, b
825w, b
620m, vb
595sh
562sh
510w, b
465w, b

Reference 10

367

WAGNERITE

$16[Mg_2(F)PO_4]$

1095sh
1075sh
1050s
650sh } b
608m } b
525b, sh
410m, vb

Reference 9

368
TRIPLITE
$8[(Mn^{II},Fe^{II})F(PO_4)]$

1095sh	}
1075–50vs, b	} vb
1020b, sh?	}
805vw	
605m, sh	} b
580m	}
455w, b	
410?	

References 9, 375

369
TRIPLOIDITE
$16[(Mn^{II},Fe^{II})_2(OH)PO_4]$

3509			595sh	} b
1090sh	}	}	570s	}
1058sh	} b	}	486w, b	
1030vs	}	}	420w	
1010sh	}	} vb		
957m, sh		}		
870b, sh		}		
755sh		}		

Reference 10

370
LIBETHENITE
$4[Cu_2(PO_4)(OH)]$

3465		555w
1635vw		522vw
1080sh		480w
1050s	} b	445w
960vs, sh	} doublet	
815w, sh		
650sh	}	
637m	} b	
618m, sh	}	

References 9, 597, 618

371
ROCKBRIDGEITE
$(Fe^{II},Mn)Fe^{III}{}_4(OH)_5(PO_4)_3$

3590		640sh	}
3250		600m, b	} b
1145sh	}	580sh	}
1060m, sh	}	555sh	}
1030s	}	440b, sh	} b
970m, b, sh	} vb	415m	}
890sh	}		
740w, b	}		

Reference 10

372
TARBUTTITE
$8[Zn_2(OH)PO_4]$

3440		612s, b	}
1600vw		590m, sh	} b
1095s, sh	}	526m, sh	}
1040m, sh	}	452sh	
1005vs	}	436m	
962m, sh	} vb		
920s, sh	}		
808m	}		

Reference 10

373
AUGELITE
$4[Al_2(OH)_3PO_4]$

3555s	} b	717m	
3465s	} doublet	640s, b	}
1205s, sh	}	615m, sh	} b
1155s	}	556s, b	}
1079vs, sh	} vb	438s, b	
1015s	}		
930vs, sh	}		
835m	}		

Reference 10

374
DUFRENITE
$Fe^{II}Fe^{III}_4(OH)_5(PO_4)_3 . 2H_2O$

3580
3280
3200
1630vw, b
1135sh }
1070b, sh }
1032vs } vb
995m, sh }
870sh }
800sh }
600b, sh } b
575m, b }
455b, sh
435m, b

Reference 10

375
APATITE/FLUORAPATITE
$2[Ca_5F(PO_4)_3]$

1100–1080s, sh
1050vs, b
1025b, sh
970–60m, sh
605s, sp } b
580s, sp } doublet
565sh
478vw
438vw
325vs, sh

References 4, 275, 425, 597–600, 612

376
CHLORAPATITE
$2[Ca_5(PO_4)_3Cl]$

1085sh }
1050vs } b
960b, sh }

References 177, 600

377
FLUORCHLORAPATITE
$Ca_{10}(PO_4)_6F_{2-x}Cl_x$?

1640vw
1460m } b
1432m } doublet
1102–1082sh
1055–1050vs } b
1025vs } doublet
970–962m, sh
610–605 } s, b
573 } doublet
555sh
476–472w, sp
430sh
ca. 335s, sh

References 3, 9, 21, 425

378

HYDROXYAPATITE

$2[Ca_5(PO_4)_3OH]$

ca. 3500w		475w
ca. 2100w, b		425w, sh
1110–1095s, sh	b	385s
1050–40vs, b		
970–45sh		
745w		
613–05vs	b	
600sh		
580sh		
576–65vs		

References 9, 36, 109, 177, 201, 257, 275, 375 (poor quality), *425, 420* (Var. Francolite), *600*

379

DAHLLITE

$2[Ca_5(PO_4,CO_3,OH)_3OH]$

1470sh		Carbonate spectrum only
1450vs	b doublet	
1410vs		

Reference 420

380

YTTRIAN APATITE

$2[(Ca,Y)_5(PO_4)_3(F,OH)]$

3500
3400
3300w, b
1090sh
1040vs
955sh
715w, b

Reference 109

381

PYROMORPHITE

$2[Pb_5(PO_4)_3Cl]$

1630vw	436vw
1440vw ?	245s
1320vw ?	171s, b
1110sh	148s, b
1030s, sh	
999sh	
970–67vs	
927sh	
575s	
543s	

References 3, 9, 21, 116, 257, 432, 509, 533 (Xtal-reflection), *597, 598–600*

382
LAZULITE
$2[(Mg,Fe^{II})Al_2(PO_4)_2(OH)_2]$

3550	
3460	
3400	
1630vw	
1200–1190m, sh	vb
1150sh	vb
1115sh	vb
1085–80vs	vb
1035m, sh	vb
1010–00m, b, sh	vb
930w, sh	
820–12m, b, sh	b doublet
775–70m	b doublet
652s	vb
590sh	vb
555s, b	vb
512s, b	vb
470s, b	vb
430m, sh	vb

References 9, 376 Barbosalite *3*

383
TSUMEBITE
$Pb_2Cu(OH)_3(PO_4) . 3H_2O$

3440		630sh	b quartet
1635w		615m, sh	b quartet
1090vs	vb	596sh	b quartet
1048sh	vb	575m, sh	b quartet
1015sh	vb	515sh	
970s, sh	vb	475vw	
932sh	vb	425vw	
800b, sh	vb	415vw	

Reference 9

384
TURQUOISE/KALLAITE
$[CuAl_6(OH)_8(PO_4)_4 \cdot 5H_2O]$

3530–08
3463–60
3446
3300–3295
3090
1650–40m, b
1195sh
1175vs
1115vs } b quartet (1195–1016)
1063vs
1016s, sp, sh
905vw
838m, sp
790m, b, sh
695sh
655s, b
615s } partially
595s } resolved } vb (790–460)
580s } triplet
555s
480s
460m, sh
435vw, b
415vw, b

References 9, 111, 116

385
LUDLAMITE
$2[(Fe^{II},Mg,Mn)_3(PO_4)_2 \cdot 4H_2O]$

3170
1592m
1095sh
1070inf
1030vs } b
995vs } doublet } vb (1095–920)
940vs, sh
920sh
800vw
760vw
705vw, b
662vw
635sh
600m, sh
575s } b (635–540)
560sh
540sh
492w
465vw

References 3, 10

386
BERAUNITE/ELEONORITE
$Fe^{II}Fe^{III}_4(PO_4)_3(OH)_5 . 3H_2O$

3580–60sh
3500–3470sp, sh
3380?
3250?
1635–13w
1155–40m, sh
1110–00sh
1076–60sh
1035–26vs
990–84m, sh
965–56m, sh
930–25m, sh
890sh
805–795w, b, sh
(1155–40m, sh through 805–795w, b, sh: vb)
682–75m, vb, sh
585–77s, vb
510sh
483–70s, vb
448b, sh
413b, sh

References 10, 108

387
TINTICITE
$Fe_3(PO_4)_2(OH)_3 . 3\frac{1}{2}H_2O$

3820sp, sh
3350vs, b
1620m, b
1127vs
ca. 1070sh
1013vs
958s, sh
873m, sh
(1127vs through 873m, sh: b)
740m
628s, sh
583sh
552s
(628s, sh through 552s: b)

Reference 108

388
WAVELLITE
$2[Al_6(PO_4)_4(OH)_6 . 9H_2O]$

3555–20m, sp
3440
3280–40
3185–00w, b
1695?
1645–07m, b
1170–45sh
1110–1090sh
1065–44vs
1040–14sh
940–35sh
863–46w
660–42s, b
602–586m, b, sh
557–45m, b, sh
500s
452s
(660–42s, b through 452s: vb)

References 4 (almost complete disagreement with above), *9, 111, 375, 376*

389
TORBERNITE
$[Cu(UO_2)_2(PO_4)_2 . 12H_2O]$

3570
3430
2940
1635w, b
1410b, sh
1115b, sh
1013vs
915m, sh
805vw
695w, b
615w, b
550m
465w, b
(805vw through 465w, b: vb)

References 9, 116

390
AUTUNITE
$8[Ca(UO_2)_2(PO_4)_2 . 12H_2O]$

3435	
2935	
ca. 1645m, b	
ca. 1410w, b	
1123m, b, sh	
1023vs	
920s, sh	
690sh	vb
650sh	vb
610sh	vb
545s	vb
472w	vb
435s	vb

References 3, 9, 21, 116

391
URANOCIRCITE
$Ba(UO_2)_2(PO_4)_2 . 8H_2O$

3430	
2935	
1630m, b	
1410vw	
1115sh	b
1023vs	b
913s, sh	b
802w	b
675sh	vb
620sh	vb
545s	vb
472sh	vb
430sh	vb

References 9, 116

392
HAGENDORFITE
$(Na,Ca)(Fe^{II},Mn^{II})_2(PO_4)_2$

1080sh	vb	465vw
1000vs, b	vb	420w
960sh	vb	
645sh	b	
600sh	b	
575s	b	
550s, sh	b	
520sh	b	

Reference 10

393
HURLBUTITE
$4[CaBe_2(PO_4)_2]$

1233s, sh	789sh	546m
1164sh	771sh	524vw
1119vs	760s, sp	510sh
1072vs	724sp, sh	500m
1043vs	694s	
1008vs	618w	
853sh	592m, sh	
840w	575s	

References 172, 602

394
CERIUM PHOSPHOROCHATTONITE
$[Th_{0.5}(Ce,La,Nd)_{0.4}(U,Fe)_{0.1}]$
$(Si_{0.5}P_{0.5})(O_{3.4}OH_{0.5}F_{0.1})_4$

1120sh } b
1000s }
880sh
625w }
565w } b
545w }
460m }

Reference 394

395
NINGYOITE
$(U,Ca,Ce)_2(PO_4)_2 \cdot 1\text{–}2H_2O$

1620w
1050s, vb   Poor spectrum

Reference 101

396
SCHODERITE
$Al_2(VO_4)_2(PO_4)_2 \cdot 6H_2O$

3700vs
3330sh
1665m
1150s, b, sh
1040vs, b
960s, b, sh
895w
835w, b
740w, b

Reference 106

397
BASIC MAGNESIUM PHOSPHATE (*S*)
$Mg_4(PO_4)_2(OH)_2$

1045vs
970m, sh
845m
735m
660w
585m } b doublet
555m }
505w }
490w }
470w } b quintet
455w }
430w }

Reference 4

OTHER MINERAL PHOSPHATES

Arlidge and Farmer[376] have included some phosphate mineral spectra from which the absorption positions cannot be confidently calculated.

Synthetic Metavariscite, $AlPO_4 . 2H_2O$
Synthetic Leucophosphites
Synthetic Phosphotridymite—Taranakites
Lead Hydroxyapatite, Lead Oxyapatite are given by Engel[649].

OTHER PHOSPHATES

Monobasic	*References*	*Dibasic*	*References*
NH_4	*4, 7, 32, 47*	NH_4	*4, 7, 32, 47, 152*
Mg	*7, 32*	Ba	*4, 7, 32*
K	*4, 7, 32, 47, 418*(ATR)	K	*4, 7, 32, 47, 76, 418*(ATR)
Ca	*7, 32, 47, 152, 153, 179*	Ca	*4, 47, 152, 153*
Na	*7, 32, 47, 256, 308, 418*(ATR)	Na	*7, 32, 47, 256, 308, 418*(ATR)

Tribasic

Nyquist and Kagel[4] have included the following tribasic phosphates:
B, Sb, Bi, Ag, Hg (NH_4,Mn), (Pb,Cu), Al, Sn, Cd, Pb,Li, Fe^{II}, Fe^{III}, (NH_4Co), (Li_2,Na), (Ca,Ni), Cr, Cu, Ni, Mg, Na

Anhydrous	*References*	*Hydrated*	*References*
K	*7, 32, 418*(ATR)	Na	*7, 32, 256, 418*(ATR)
Pb	*7, 32, 47*	Cr	*7, 32, 47*
B	*150, 331*	Cu	*7, 32*
Cd	*150, 315*	Ni	*7, 32*
Bi	*150, 315*	Mg	*7, 32, 47*
La	*435*	Mn	*7, 32*
Zn	*76*	Ga	*331, 440*
Tl	*76*	Co	*315*
		Ba	*47, 150, 315, 435, 447*
		Sr	*47, 315*
		Ag	*76, 435*

PERCHLORATES

Miller *et al.*[7,32] give the spectra of the following perchlorates:
Mg, K, Na, NH_4.

Nyquist and Kagel[4] have included the following perchlorates:
NH_4, Ba, Ce, Cs, Ga, Li, Mg, Rb, Na, Zn.

PERIODATE

398
POTASSIUM PERIODATE (*S*)
KIO_4

1640w	352w, sh
858s, sh	335s, sh
850s	327sh
690vw	317vs
617w	265s
597vw	140s, vb
570vw	
555vw	
470w, b	

References 4, 7, 32

PERRHENATE

399
POTASSIUM PERRHENATE (*S*)
$KReO_4$

930sh
913vs
898–870vs
361m
317s
304s
265m

References 323, 396

PERMANGANATES

Apart from the permanganates given by Nyquist and Kagel[4] (Li, Na, K, Mg, Zn, Ba) the following may also be found:

	References
Na	*7, 32*
K	*7, 32, 396*
Ca	*7*
Ba	*7, 32, 396, 412, 447*

TUNGSTATES

400
HUEBNERITE
[$MnWO_4$]

880–72m, sh
840–20s, b
712–10s, b, sh
625–20vs, b
525–20s, sh
462–50m, sh
426–20m, sh
330s, vb

(lines 880–330 bracketed: vb)

References 9, 116, 305, 423, 564, 572, 611

401
WOLFRAMITE
2[$(Mn,Fe)WO_4$]

872m, sh
823vs
705b, sh
625vs, b
530m, b
465m, b

(lines 872–465 bracketed: vb)

425 var. intensity

References 9, 116

402
FERBERITE
2[$FeWO_4$]

870sh
818s, b
700–660vb, sh
625s, b
535b, sh
470sh
425sh

(lines 870–425 bracketed: vb)

References 9, 116, 564, 611

403
SCHEELITE (*N* and *S*)
4[$CaWO_4$]

3450w, b?
1625w?
1120vw
1020vw
870–65sh
825–794vs, b
442–40w
330–25s, b
285m

References 3, 7, 9, 32, 305, 419 (Xtals), *423, 533* (Xtal-reflection), *564*

404

STOLZITE

4[$PbWO_4$]

850–40sh } b
782–70vs }
375m, sp
355

References 9, 419 (Xtals), *423, 564, 572*

405

CADMIUM TUNGSTATE (*S*)

$CdWO_4$

885–80sh }
850–20s }
700b, sh }
620–00vs, b } vb
560? }
515–05m, sh }
460–55m, sh }
410m, sh }
355–50sh
325–10m, vb

References 4, 305, 423 (Xtals), *572*

406

SODIUM TUNGSTATE (*S*)

$Na_2WO_4 . 2H_2O$

3450
3330–10vs
3280–50sh
2250w } doublet
2230w }
1863vw
1820vw
1790vw
1700–1695sh
1686–70m
1640sh
933–25w, sp, sh
860–50s, sh }
835–22vs } b
827–08sh }
725vw
685–80sh }
645m } b
605sh }
548–45b }

References 4, 7, 10, 32, 225, 305, 323, 423

OTHER TUNGSTATES

Clark and Doyle[423] have recorded the infrared spectra of several tungstates:
Co, Ni, Mg, Zn, Na, Rb, Cs, Cr, Cu, Ag, Sr, Ba, K, Li.

Nyquist and Kagel[4] have included the following tungstates:
Mg, Sr, Ba, Al, Cu, Zn, Ag, K, Li.

Zorina and Syritso[572]:
Sr, Ba, Mg, Cu, Ni, Co, Zn.

Minerals		*References*
Russellite		*564*
Ferritungstite	$n.Fe_2HO_6 . 6H_2O$	*564*

MOLYBDATES

407
POWELLITE (*S*)
$4[CaMoO_4]$

3450w
1640w
1080vw
840vs, b, sh
805–00vs, b
430–28m, sp
328–25s, b
290sh
230s, b
155s, sh

References 4, 315, 419 (Xtals), *423, 564*

408
WULFENITE
$4[PbMoO_4]$

1630vw		378–70m
1540vw		350m
1440vw		310–05s
870sh	vb	270–65m
855sh	vb	
840–30sh	vb	
785–75vs	vb	
700m, vb, sh	vb	
590vw		
510vw		

References 3, 4, 9, 116, 419 (Xtals), *423, 564*

409
AMMONIUM
HEPTAMOLYBDATE (*S*)
$(NH_4)_6Mo_7O_{24}\,.\,4H_2O$

3450sh	640vs, b
3130vs	580sh
3030sh	475m, b
2850sh	414w } doublet
1640m	405w } doublet
1410vs	385sh
925–10sh	372m
885vs, b	355
840sh	

References 4, 7, 32, 205

OTHER MOLYBDATES

Clark and Doyle[423] have recorded the infrared spectra of several molybdates:
Ag, Sr, Ba, Cd, Zr, Rb, Cs, Hf, VO, Cr, Cu, Zn, Ni, Co, Li, K, Na.
Nyquist and Kagel[4] have included the following molybdates:
Ag, Sr, Ba, Cd, Cu, Zn, Zr, Ni, Co, Li, K, Na.

Minerals		*References*
Ferrimolybdite		*564, 632, 634*
Koechlinite	$4[(BiO)_2MoO_4]$	*564*
Umohoite		*564*

CHROMATES

410
ALUMINIUM CHROMATE (*S*)
$Al_2(CrO_4)_3$

3520 } s, b	560w } b
3460–50 } s, b	520b } b
1640–25w	455w } b
1130w, sh } b	425w } b
1035inf } b	410 } b
1020–1010s } b	380sh
950s, b } b	370s
930sh } b	350sh
800w, sh	
745–30s, b	

References 4, 7, 32

411
AMMONIUM CHROMATE (*S*)
$(NH_4)_2CrO_4$

3230s, sh	850–45w, sh
2980s	787w, b
2810s, sh	760–42w, sh
1640w	
1425–20s } doublet?	
1417–10s } doublet?	
947–40m, sh	
925–20m, sh	
875–62s, b	

References 4, 7, 32, 151

412
BARIUM CHROMATE (*S*)
$BaCrO_4$

3450w
944–30s, sh
890–85 } vs, b doublet?
850–43 } vs, b doublet?
420m
392m
375w

References 7, 32, 437

413
CADMIUM CHROMATE (*S*)
$CdCrO_4$

α	*β*
915vs } b	970vs
865s, sh } b	940s, sh
805vs, b } b	915vs, sp, sh
420m	835s } b doublet
385m, sp	820s } b doublet
365sh	715s
	440m
	390w, sh
	365w, sh
	355sp, sh
	335w

References 4, 315, 443 (α and β)

414
CALCIUM CHROMATE (*S*)
$CaCrO_4$

ca. 3550s
1660m
990sh?
890–70vs, b
345m, sh
260s, vb
180s, sh

References 4, 315

415
CROCOITE
$4[PbCrO_4]$

905–895sh }
858–50vs } b
835–30sh }
720w, sh }
425vw
340w

References 4, 7, 9, 116

416
MAGNESIUM CHROMATE (*S*)
$MgCrO_4 . 7H_2O$

3540sh
3260vs, b
2270m, b
1660–50sh } doublet
1630–20s }
925inf }
910sh? }
894b, sh } b
880–75vs }
855m, sh }
765w, b
695w, vb

References 4, 7, 443

417
TARAPACAITE (*S*)
$4[K_2CrO_4]$

946–40sh }
920inf } b
889–75vs }
858–52sh }
720m, sh
400sp } w doublet
387sp }

References 4, 7, 10, 32, 437

418

SODIUM CHROMATE (*S*)

$Na_2CrO_4 . xH_2O$

3350s, b
3170s, b
2190–50w, b
1680–66m
910sh ⎫
890–85s ⎪
865–60sh ⎬ b
854–45m–s ⎪
820? ⎭
720w, sh
630–10m

References 4, 7, 32

419

ZINC CHROMATE (*S*)

$ZnCrO_4 . 7H_2O$

3450s, b
2700w, b
1218–1185m ⎫
ca. 1090s ⎬ b
1055–50sh ⎭
942–40s, sh ⎫ b
ca. 875vs ⎭
810–795vs, b
720–15m, b
638w
612w, b
410w, b
385w, b

References 7, 32

OTHER CHROMATES

	References
$CoCrO_4$	*315*
$MgCrO_4$	*463*
$Ba_3(CrO_4)_2$	*412, 447*
$Li_2CrO_4 . xH_2O$	*4, 447*
$CsCrO_4$	*4*

DICHROMATES

420
AMMONIUM DICHROMATE (*S*)
$(NH_4)_2Cr_2O_7$

3220–3180s, sh
3075–00b, sh
2820–00sh
2320
1650
1410vs
950vs }
930sh } b
900s, sh }
875s, sh }
800s, sh } b
ca. 740s, b }
577w
370s

References 4, 7, 32

421
CALCIUM DICHROMATE (*S*)
$CaCr_2O_7 . 3H_2O$

ca. 3500m
1650sh } intensities may be reversed
1620m }
970m, sh
940vs }
900–890s, sh } b
820–10s, b }
760 ?
720–10sh
567w }
540w } b
520w }
400vw
375w, sh
365w, sh
359m, b

References 4, 7, 32

422
LOPEZITE (*S*)
$4[K_2Cr_2O_7]$

1310w
967sh
935sh
925sh
910sh } doublet ?
905s, sp }
890sh } doublet ?
885s, sp }
800m, sh
ca. 760s, b
595w
567w } doublet
555w }
470–62w
392w, sh
385sh
378m
330w

References 4, 7, 10, 32, 586

423
SODIUM DICHROMATE (*S*)
$Na_2Cr_2O_7 . 2H_2O$

ca. 3500s
1655s }
1625s } triplet ?
1615s }
952sh
940vs, b
910m
890s, sh
780sh
740vs, b
630w }
575w } vb
533w }
480w }
410w
375sh
367m
350w

References 4, 7, 32

OTHER DICHROMATES

	References
$CuCr_2O_7,2 . H_2O$	*7, 32*
$Ag_2Cr_2O_7$	*4*
$ZnCr_2O_7$	*4*
$Rb_2Cr_2O_7$	*4, 586*
$Li_2Cr_2O_7 . 2H_2O$	*4*
$Cs_2Cr_2O_7$	*586*

GERMANATES

Monticellite, olivine and spinel structure germanates are discussed in detail in Refs. 44, 121, 122.

	References
Co_2GeO_4	*339*
Ni_2GeO_4	*44, 169*
Zn_2GeO_4	*75*
Na_2GeO_4	*79*
$Mg(Ge_4O_{10})(OH)_2$	*398*

424
PYROPE
$8[Mg_3Al_2Si_3O_{12}]$

2900?
1080–70sh
1010–995sh
980–63vs
910–894vs } b doublet
880–66vs }
820–15sh
580–75s, b
530sh
485–80vs } b doublet
465–60vs }

References 9, 127, 183, 202, 612

425
ALMANDINE
$8[Fe^{III}{}_3Al_2Si_3O_{12}]$

1170w, sh }
1090m, sh }
1007–997sh }
970–63vs } b
905–896vs } doublet }
880–72vs } doublet }
855w, sh }
637w
570s, b
530–27sh
480s, b
455sh

References 7, 9, 17 (reflection), *119, 183, 202, 533* (Xtal-reflection)

426
SPESSARTINE
$8[Mn_3Al_2Si_3O_{12}]$

1070w, b, sh
1000–980sh
955–50vs
892–84vs } b doublet
870–64vs }
855sh
632–30w
565–62s, b
520sh
472–70vs } b doublet
452–50vs }

References 9, 122, 183, 202, 335, 452 (poor)

427
UVAROVITE
$8[Ca_3Cr_2Si_3O_{12}]$

1080sh
945–40sh
900–887vs
840vs?
825–20 } doublet
819–15 }
610m
540vs }
455vs } b
425vs }
375s }

References 9, 122, 183, 202, 335

428
GROSSULAR
[$Ca_3Al_2Si_3O_{12}$]

2800?
1080sh
960–50sh
915–07vs
860–52vs } b doublet ?
840–33vs }
618–15m
550inf
540s
515inf
470m, sh
450vs
400s, sp

References 9, 84, 122, 183, 202, 335, 533 (Xtal-reflection)

429
ANDRADITE
8[$Ca_3Fe^{III}{}_2Si_3O_{12}$]

ca. 1500?
1090–85b, sh
945–30sh
895–87vs, sp } b doublet
840–26vs, sp }
820–13sp, sh
640vw?
592–90w
512–10s
482–80m
440–38s

References 9, 119, 122, 183, 202, 226, 626

430
SCHORLOMITE
[$Ca_3(Fe,Ti)_2(Si,Ti)_3O_{12}$]

940sh }
890vs } b doublet
840vs }
820sh }
740sh
595w }
520s }
480s, sh }
440vs, sh } vb
400vs }
360s, sh }
330vs, sh }

References 43, 335

431
IVAARITE
[$Ca_3(Fe,Ti)_2(Si,Ti)_3O_{12}$]

910m, vb, sh
830vs, b
660m, vb
580m, sh
520m, sh
435s, sh } b doublet
390vs }
350m, sh
320m, sh

References 43, 335

432
APLOME/POLYADELPHITE
$8[(Ca,Mn)_3Fe_2Si_3O_{12}]$

1180b, sh ⎫
955sh ⎫ ⎪
912vs ⎬ b doublet ⎬ vb
850vs, b ⎪ ⎪
830sh ⎭ ⎪
685b, sh ⎭
610w ⎫
525s, b ⎪
486sh ⎬ vb
448s ⎪
427sh ⎭

Reference 10

433
MELANITE
$8[Ca_3(Fe,Ti)_2(Si,Ti)_3O_{12}]$

1080w, sh
950sh
905vs
860sh
840vs, b
750sh
650w, b
600w
520s
485sh
447vs
425sh

References 9, 43, 202

434
EULYTINE
$Bi_4Si_3O_{12}$

ca. 1100vw, vb
980–60sh
900–890vs
808–00m, sh
550–42m
500–485m–s

References 9, 119

435
ZIRCON
$4[ZrSiO_4]$

ca. 3400w, b?
1020–00sh
900–895vs, b
618–00s, sp
455–40sh (doublet?)
400s
320s

References 9, 103, 119, 301, 373, 465, 587, 589, 638, also Malacon, Alvite and Naegite *589*

436
CYRTOLITE
$ZrSiO_4$?

980–20vs, b
ca. 800w, sh
620w
535m, b
440w?

References 373, 587, 589

437
THORITE
$ThSiO_4$

1650w, b
ca. 1200sh
1010–990vs, vb ⎫
900sh ⎬ b
ca. 700w? ⎭
460s, vb
410sh

References 10, 103

438
PHENAKITE
$6[Be_2SiO_4]$

1075sh?		b (1075–901)
1044–40vs	v triplet	
945–41vs	v triplet	
914–08s, sh	v triplet	
901vs		
820–10b, sh	b	
780–75vs	b	
750–44sh	b	
735s	b	
720–15vs	b	
698–2s	b	
645–37m, sh		
600–595s		
580–78sh		
427s		

References 9, 119, 172

439
WILLEMITE
$6[Zn_2SiO_4]$

1080b, sh	
980–70s, sh	
935–29vs	doublet
905–00vs	doublet
885–68s	
852sh	
620–13s, sh	
600sh	
580–75s, sp	
465–60s, sp	
400–396	s doublet
385–80	s doublet

References 3, 4, 9, 75, 80, 119, 301, 500, 533 (Xtal-reflection)

440
TROOSTITE
$6[(Zn,Mn)_2SiO_4]$

(approx.)	
1000s, sh	
940vs	doublet?
910vs	doublet?
890sh	
615m, b	
585m, b	

Reference 17

441
OLIVINE
$4[(Mg,Fe)_2SiO_4]$

1100b, sh
1002s, sh
950sh
898s
848sh
612m
516s, b
475b, sh
418s

References 9, 533 (Xtal-reflection)

442
LARNITE (*S*)
$4[\beta\text{-}Ca_2SiO_4]$

ca. 1490vw, b	
ca. 1430vw, b	
1035inf	very broad band with variable intensity of the individual absorption bands
1000–996m, b, sh	
939–35sh	
922sh	
905–00s, b, sh	
872–70sh	
850–48vs	
540sh	
523–21vs, b	
440w	

References 4, 10, 80

443
CALCIO-OLIVINE (*S*)
$4[\gamma\text{-}Ca_2SiO_4]$

950s, b, sh	
940sh	
925sp, sh	
860vs	
820m, sh	
570m	
520s	b doublet
500s	
440w, sp	
390m, sh	
340vs	
310m, sh	

References 10, 80, 121, 261

444
MERWINITE
$Ca_3Mg(SiO_4)_2$

1025s	b
995s	
975m, sh	
945vs, b	
900s, sh	
890vs	
870s, sh	

Reference 261

445
AXINITE
$2[Ca_2(Mn,Fe)(Al_2B)Si_4O_{15}(OH)]$

1108–03vs	var. intensities may be reversed	vb
1075–66sh		
1052–44vs		
1015–09sh		
988vs	not always resolved	
935sh		
890m, sh	doublet?	
884m, sh		
785m, sh		
667s, sh	vb	
635sh		
605inf		
590vs		
555vs		
521s, sh		
483m, sh		
460m, sh		
437m, sh		
412m, sh		

References 9, 18 (reflection), *50, 120*

446
FORSTERITE
$4[Mg_2SiO_4]$

1000–990s	b
965–58sh	b
890vs, b	b
840m, sh	b
620–02s	
545sh	
524sh	
512–01vs	
473–63sh	
430–28vs	
409–05s	

References 1, 3, 9, 21, 119, 121, 227, 281, 592

447
FAYALITE
$4[Fe_2SiO_4]$ (Fa_{100})

1060w, b, sh	b
965–55sh	b
950–47s, sh	b
920–15sh	b
880–73vs	b
828m, sh	b
566–58m, b, sh	
510–02sh	
495sh	
482–80m–s, b	
370m, sh	
320m	

References 1, 10, 121, 281, 301, 592

448
HORTONOLITE
$4[(Fe,Mg)_2SiO_4]$ $(Fo_{40}Fa_{60})$

1080sh	b
970sh	b
955s, sh	b
920sh	b
875vs (840?)	b
825sp, sh (780?)	b
573–60s, b, sh	b
498–90s, vb	
410w	

References 9, 119

449
KNEBELITE
$4[(Mn,Fe)_2SiO_4]$

1058sh	vb
1020sh	vb
953–50s, sh	vb
940sh	vb
912inf	vb
880–70m, sp, sh	vb
725sh	vb
663vw	
565–60vs, sh	
505sh	
475s, vb	

References 10, 281

450
TEPHROITE
$4[Mn_2SiO_4]$

978–75sh	variable intensity vb
968–43s, sh	
935–08sh	
875–55vs	
826–13s, sp, sh	
585–70s	
510–08b, sh	
485s	
435sh	
360s, sh	
300vs	

References 10, 121, 452 (poor)

451
MONTICELLITE
$4[CaMgSiO_4]$

1060b, sh	vb
982–71vs	
955–42m, sh	
890–80vs, b	
833–25s, sh	
597–87m	variable intensity
520–17s	
485–80sh	
432–25s	b doublet
382s	
311s	

References 9, 119, 121, 261, 281, 301

452
GLAUCOCHROITE
$4[(CaMn)_2SiO_4]$

956s	b
938sh	
915sh	
865vs	
816m, sh	
566s	

Reference 281

453
HILLEBRANDITE
$12[Ca_2SiO_3(OH)_2]$

3635		796vw	
3616		748vw	
3592		700sh	b
3556		650m, b	
1610w		530w, sh	
1445w		512sh	
1076vs	b quartet	453s, b	
1034vs		405vw	
1020sh			
990sh			
966vs			
935sh ?			
906vs			

References 10, 421

454

CERITE

$(Ca,Ce)_3Si_2(O,OH,F)_9$

1450w	
975sh	
955m, sh	
918sh	vb
885s, b	
820sh	
690b, sh	
545sh	
528m, sh	b
500b, sh	
485m	
422vw	

Reference 10

455

VESUVIANITE/IDOCRASE

$4[Ca_{10}(Mg,Fe)_2(OH)_4Al_4Si_9O_{34}]$

3570		
3440		
1640vw		
1030–12vs	b	
995–63vs	triplet	
925–16s, sh		vb
910sh		
879–70sh		
805–797m, sh		
691w, sh	doublet	
680w, sh		
640–31sh		
608–00sh		
580–66s, sh		
540–24sh	vb	
495–80s		
450–40sh		
418–10m, sh		

References 1, 9, 36, 397

456

TRITOMITE

$Ca_3(La,Ce)_3Zr_3(F)_6(BO_3)_3(Si_3O_9)_2$

1655vw	
1330vb, sh	
1200vb, sh	vb
970s, vb	
730w	
475m, vb	

Reference 10

457

STEENSTRUPINE

$Na_2Ce(Mn,Fe,Ta,...)H_2((Si,P)C_4)_3$?

1655vw
1030vs, vb
940b, sh
612sh
580b, sh
540b, sh
455m, vb

Reference 10

458
TOPAZ
$4[AlSiO_4(OH,F)_2]$

4550vw		Intensities depend on crystal thickness
3850w		
3570w	doublet	
3330w		
2325vw		
1850w		
1171s, sh		
1080b, sh		
1010m, sh		
960–50s	b	
935sh		
900–880s		
775sh		
708m, sh		
665sh	b	
655–50vs		
635–20vs		
585sh		
562–50m	b	
530–20m, b, sh		
483m, sh		
445vs		
425sh		

References 3, 9, 17, 18 (reflection), *21, 347, 533* (Xtal-reflection), *631*

459
ILVAITE
$4[CaFe^{II}{}_2Fe^{III}(OH)Si_2O_8]$

ca. 3400
1038–32s, sh
982sh
958–45m, sh
905–01vs
823–18s, sh
765–54sh
707–697w, sp
573–70m, sh
540–36m, sh
496–94s
452–50m, b, sh
432sh

References 9, 80, 500

460
ROEBLINGITE
$Ca_7Pb_2Si_6O_{14}(OH)_{10}(SO_4)_2$

3160		788vw
1312w		764vw
1250vw		742vw
1200vw	}	707vw
1170sh	}	675vw
1090m, sh	}	605m
1040vs	}	576m
1005sh	} vb	550sh
965s, sh	}	513w
925sh	}	472w, b
865vs, sh	} b? }	445vw
850sh	} b? }	423vw
		415vw

Reference 10

461
NORBERGITE
$4[Mg_3(OH,F)_2SiO_4]$

3560		630m, b	}
3530		567m	}
1002s, sh	}	553m, sh	} vb
965sh	}	488sh	}
925sh	} b	440m, sh	}
902s	}	406s, vb	}
886sh	}		
865sh	}		
756w			

Reference 10

462
CHONDRODITE
$2[Mg_5Si_2O_8(F,OH)_2]$

1000s, sh	}
920sh	} b
900s	}
855m, sh	}
758w	
608m, b	
555m, b	
480m, b	
440inf?	
412s, b	

References 9, 18 (reflection)

463
HUMITE
$4[Mg_7(OH,F)_2(SiO_4)_3]$

1000s	}
960m, sh	} b
892vs	}
852m, sh	}
760w	
618m	
540m, b	
485s, b, sh	
415s, b	

Reference 9

464
WOEHLERITE
$4[NaCa_2(Zr,Nb)Si_2O_8(O,OH,F)]$

1070m, sh		730sh
1060sh		670b, sh
1025s, sh		628m, b
1000sh		585sh
960sh	vb	560m
915m, sh		537sh
870s		525sh
820m, sh		477m, b
762m		450m, sh
		434b, sh

Reference 10

465
MOSANDRITE
$NaCa_6Ce_2(TiZr)_2Si_7O_{24}(OH,F)_7$

3570		800vw	
3540		785vw	
3320		722m, sh	
1645w		690sh	vb
1180sh		678m, sh	
1092m, sh		625m	
1068s	b	600sh	
1050sh		577m, sh	
990vs		540sh	
		512m	b
		487m	doublet
		443m, sh	
		420m	

Reference 10

466
RINKITE
$[(Na,Ca)_{12}(Ce,Ti)_5Si_8(O,F)_{36}]$

1060inf	
980s, b	b
885s, sh	
662vw	
592b, sh	
560b, sh	
515inf	
485m, b	
405m, sh	

Reference 10

467
HELLANDITE
$(Ca,Y,Er,Mn)_{>3}(Al,Fe)(SiO_4)_2 . H_2O$?

3430		665vw
1635vw, b		595m, b, sh
1430vw, b		540sh
1158s, b, sh		523m, vb
1050sh		515sh
980sh	vb	460m, b, sh
905vs, vb		425m, sh
746m, sh		

Reference 10

468

GUARINITE

$Ca_2NaZr(F)(SiO_4)_2$?

1075m, b, sh	vb	575m, sh	vb
1000vs, b	vb	538m, sh	vb
965sh	vb	510sh	vb
920m, sh	vb	473s, b	vb
864s, sh	vb	436m, sh	vb
737vw		405m, sh ?	vb
711vw			
663vw			

Reference 10

469

SILLIMANITE

$4[Al_2SiO_5]$

3367 ?		830–20s	
1210–00sh	vb	804	
1185–80s	vb	751–50m	
1169	vb	695–90s	b
1125	vb	640–35m, b	b
1075–70sh	vb	581–75vb, sh	b
1050–40s	vb	545–43s	b
1020–03s	vb	534sh	b
970–60vs	vb	515–08s	b
920–13vs	vb	491s, b	b
909	vb	440vs	b
895–88s, sh	vb	374w	
		346w	
		335w	
		322w	

References 3, 9, 119, 191, 264, 295, 301, 334, 459, 500, 585

470

ANDALUSITE

$4[Al_2SiO_5]$

1020–10b, sh	var. intensity; b	485–80s
980–76s	var. intensity; b	455vs
940vs	b	440sh
895–85s, sh	b	415sh
855sh		390m
780–75m, sh	b	360m
740–35m, sh	b	310s
685s, b	b	
610–00vs, b	b	
560–58sh	b	
520s	b	

References 3, 9, 36, 191, 301, 334, 500, 585

471

STAUROLITE

$[(Fe,Mg)_4Al_{18}Si_8O_{46}(OH)_2]$

3420s, b		695b, sh
ca. 1630vw, b		650s, vb
1190–70inf	b	595b, sh
1110–1090s, b	b	485s, vb
1030–990vs, vb	b	435m, sh
910sh	b	
855–50vb, sh	b	
802–00w	b doublet	
785–80w	b doublet	

References 9, 36, 301

472
KENTROLITE
$Pb_2Mn^{III}{}_2Si_2O_9$

995m, sh
932s
894m, sh
870sh
836sh
790w, sh
695w
605b, sh
525m } b, sh
505m }
450m
418sh

Reference 9

473
KORNERUPINE
$[(Mg,Fe^{II},Fe^{III},Al)_{40}(Si,B)_{18}O_{86}]$

1640vw	655m, b, sh
1170b, sh	615vs, b
1105vs	550sh
1010vs, b	520vs } b triplet
928sh	490vs }
890s	465vs }
805m, b, sh	415s, sh
740vs	
720m, sh	

All of the peaks are observed over and above a broad background.

References 9, 607

474
MELANOTEKITE
$Pb_2Fe^{III}{}_2Si_2O_9$

1650vw
1000s, sh }
938s, sh }
895s } b
838s, sh }
810sh }
694w
570b, sh }
495sh }
465s, vb } vb
410m, sh }

Reference 10

475
SPHENE/TITANITE
$4[CaTiSiO_5]$

1100sh }
960–50b, sh }
885–65vs, b } b
830w, sh }
610w, b, sh
570–65m, b, sh
500w, sh
440s, vb

References 9, 17 (reflection)

476
KYANITE/DISTHENE
$4[Al_2SiO_5]$

1640w		734sh	
1053–29m, sh		720b, sh	
1010–00m, sh		645s, b	
971sh	vb	630sh	
952–43vs		610vs	
905–00s, sh		598sh	vb
881vw, sh		572m, sh	
		550s, sh	
		512s, b	
		470s	
		442s, sh	

References 3, 4, 9, 36, 84, 191, 227, 301, 334, 500, 533 (Xtal-reflection), *585*

477
ARDENNITE
$2[Mn_5Al_5(As,V)O_4Si_5O_{20}(OH)_2 \; . \; 2H_2O]$

3620		675m, b
3180		635sh
3130		622m
1127	b	580m, sh
1108	doublet	557s
1022inf		476m, sh
975vs	b	445vs
922vs	doublet	
880sh		
848s, sh		
785sh		
732m		

Peaks from 1127–445 cm^{-1} are based on a broad sloping background.

Reference 10

478
KASOLITE
$4[PbUO_2SiO_4 \; . \; H_2O]$

3300w
1585w
955sh
895s
870w, sh
850plat
810s, b
750b, sh

Reference 87

479
ALLOPHANE
$Al_2SiO_5 \; . \; nH_2O$?

3425–10m, b
1610–08vw
1380vw
1206vw
1072–70sh
1002–00vs, b

References 127, 295, 438, 469

480
ALITE (*S*)
$18[Ca_3SiO_5]$

945–33b, sh } b
910–01sh } peak intensities vary
883–80vs }
812m, sp, sh
528–25vs, b
464sh
452–50s, vb

References 10, 80

481
HSINGHUALITE
$Ca_3Be_3Li_2Si_2O_{10}F_2$

3500w
3470w
1437w, b
1088sh } b
1047m, sh } doublet
1044m, sh }
1016m, sh
983s, sh
960s, sh
931sh
918vs, sp
894vs, sp
799sh
785sh
771s, sp
750m, sp
733m
696s, sp
681sh
610w
578vw
513vs, b
497sh
481m
467m

Reference 172

482
SAPPHIRINE
$8[Mg_2Al_4SiO_{10}]$

1080m, b, sh
985s
945sh
905sh
850s
755b, sh
720s
690sh
632sh
612sh
588s
562s, b
475vs, b
430sh
412sh?

All peaks are on a broad background.

References 9, 607

483
ZINC ORTHOSILICATE (*S*)
$ZnSiO_4 . xH_2O$?

1150sh
1060b
960sh
795vw
720vw
480m, vb

Reference 4

484

COPPER ORTHOSILICATE (*S*)

$CuSiO_4 . xH_2O$

3400vs, vb
1640m
1080vb
800m, b, sh
730sh
670vw
620vw
470m

Reference 4

485

COBALT ORTHOSILICATE (*S*)

$CoSiO_4$

970s, sh ⎫
945sp, sh ⎪
920sh ⎬ vs
870vs ⎪
825m, sh ⎭
570m
510sh
480m
380s, sh
340sh
320s, vb
260m, sh

Reference 4

SOROSILICATES

486
MELILITE
$2[(Ca,Na_2)(Mg,Fe^{II},Fe^{III},Al)(Si,Al)_2\text{-}O_7]$

3440		595w
1640vw		530b, sh
1070sh?	}	480s, b
1030–25vs	}	415w
985–80sh	} b	
940s, sh	}	
860–58m–s	}	
715sh		
685m, sh		
640sh		
617sh		

References 9, 36

487
GEHLENITE
$2[Ca_2Al_2SiO_7]$

3430	
1020sh	}
980vs	}
922vs	} b
880sh	}
860sh	}
815m, sh	}
715w	
675w	
527sh	
485s, b	
435m, b, sh	

References 9, 177

488
HARDYSTONITE
$2[Ca_2ZnSi_2O_7]$

1038s, b, sh	}
1015sh	}
971s	} b
916vs	}
895inf	}
838s	}
685w	
623m, b	
605b, sh	
550sh	
500b, sh	} b
456s, b	}

References 10, 80, 500, 593

489
MELIPHANITE/
MELINOPHANE
$8[(Ca,Na)_2Be(Si,Al)_2(O,F)_7]$

1090m, b, sh	}	570sh
945s, vb	}	560sh
895m, sh	}	542m, sh
816m, sh	} vb	515sh
745sh	}	495m, sh
715s	}	475m
683m, vb	}	455sh
670sh	}	415sh?

Reference 10

490
BARYSILITE
$Pb_3Si_2O_7$

1640vw
1430w, b
1085vw
938–35vs } b doublet
913–10sp, sh } b doublet
890–88vs } b doublet
845–30s, sh } b doublet
700m
635vw
610vw
552m
523m
485w } quartet
463s, sh } quartet
443m } quartet
ca. 415m } quartet

References 10, 198, 482

491
LEUKOPHANITE/
LEUCOPHANE
$4[(Ca,Na)_2BeSi_2(O,OH,F)_7]$

ca. 1660vw, b
1235sh } b
1090vs, sh } b
1030sh } b
1000m, sh } b quartet } b
975m, sh } b quartet } b
950sh } b quartet } b
931vs } b quartet } b
883s, sh } b
810sh
785m
750m, sh
715m, sh
685m, b
650sh
625sh
615sh
580b, sh
545s
507vw
473s } b doublet
443s } b doublet
415sh

Reference 10

492
RANKINITE (*S*)
$4[Ca_3Si_2O_7]$

1002sh } b
980vs } b
950vs } b
908–03s, sh } b
847s } b
652m
560sh
538s
482s, b

Reference 80

493
THORTVEITITE
$2[(Sc,Y)_2Si_2O_7]$

1175–60s, sh
920–02vs, b
877–60m, sh
600–595s, vb
510–05b, sh
425–15m, vb

References 9, 18 (reflection), *119, 301, 472* (far-IR), *500, 607, 647*

494
AENIGMATITE
$4[Na_4(Fe^{II},Fe^{III},Ti)_{13}Si_{12}O_{42}]$

1060–20vs, b
950–35vs
885–65sh
830?
790sh
730w
655sh?
642m
560–45s, b, sh
506m, sh
475s, b, sh
468–50vs, b
440sh?
415sh?

References 9, 18 (reflection)

495
AFWILLITE
$4[Ca_3(SiO_3OH)_2 . 2H_2O]$

(approx.)
3340w
3130w
2800w
2400w
1330w
1280w
1100m
1000sh
970s
910s
880b, sh
855sh
810m
780m
515s, b
475s, b

References 173, 421

496
GANOMALITE
$Ca_2Pb_3Si_3O_{11}$

1630vw
ca. 1450vw, b
1010m } vb
915s } b doublet } vb
865s } b doublet } vb
818s, sh } vb
770sh } vb
676m
567m
530sh
522m
480m
458m, sh
422sh
405?

Reference 10

497
NASONITE
$2[Pb_6Ca_4Cl_2(Si_2O_7)_3]$

1100sh } b } vb
1070s } b } vb
1045m, sh } b } vb
1008m, sh } vb
985sh } vb
940s } b doublet } vb
885s } b doublet } vb
780sh
663w
635sh
602sh
580m } b doublet
550m } b doublet
517s, sh
480m, sh
455m
430m, sh
412sh

Reference 10

498
BERTRANDITE
$4[Be_4(Si_2O_7)(OH)_2]$

3690m, sh	898s, sp, sh
3650sh	828s, sp, sh
3620m, sh	802–783sh
3585s, sp } doublet	776sh } b
3509s, sp } doublet	755sp, sh } b
1635vw	736–34sh } b
1390w	719–16sh } b
1170sh	695–91m } b
1133sh	675sh? } b
1122–08m, sh	637w
1040–30s	610w
1015sh	546sh
987sh	542–40m–s
956–45m, sh	524w
936–20sh	485–73w–s?
	432s, b

References 9, 172

499
CUSPIDINE
$Ca_4Si_2O_7(F_2OH)_2$

ca. 3550w } From synthetic material
ca. 2950w } From synthetic material
2330vw } From synthetic material
1425vw, b } vb
1125sh } vb
1065m, sh } vb
1038m, sh } vb
970vs, b } vb
915sh } vb
860s } vb
658w } vb
555sh } b } vb
544s } b } vb
517m, sh } b } vb
495s, b } vb
440b, sh } vb
408vw } vb

The synthetic material gave absorption bands in almost identical positions.

References 10, 260

500
MURMANITE
$Na_2(Ti,Nb)_2Si_2O_9 . nH_2O$

3380
1645w, b
1030b, sh
940vs, vb
700vw, b
535m, vb
450b, sh
425?
410?

Reference 10

501
HEMIMORPHITE
$2[Zn_4Si_2O_7(OH)_2 . H_2O]$

3450	562s
ca. 3200w, b	545–42sh
1635–04w	498vw, sh
1090–86s, sp	450s
1031–25sh	
945–35vs	
865–62s, sp	
680–72m	
610–05s, b	

References 3, 9, 18 (reflection), *36, 80, 119, 301, 500*

502
LAWSONITE
$4[CaAl_2(OH)_4(Si_2O_6)]$

1115s, sh	557m, sh ⎫
1050sh	520b, sh ⎪
1008vs	487s ⎬ b
895sh	450sh ⎪
793w, sp	415sh ⎭
770vw	
735vw	
698w	
678m, sh	
650b, sh	

References 1, 3, 9 (not good agreement between these)

503
ZUNYITE
$4[Al_{13}Si_5O_{20}(OH,F)_{18}Cl]$

3623m, sp	540sh
3338vs, b	480vs
3200sh	462s, sh
1180s ⎫	410sh?
1140s ⎪	
1076s ⎬ b quintet	
1000s ⎪	
900s ⎭	
745sh ⎫	
725vs ⎬ b	
680sh ⎪	
655vs, vb ⎭	

Reference 10

504
BATISITE
$Na_2BaTi_2(Si_2O_7)_2$

1189b, sh ⎫	573sh ⎫
1124s ⎪	531s ⎪
1050s ⎬ vb	509s ⎬ b
998sh ⎪	492s ⎪
982m, sh ⎪	469sh ⎭
961vs ⎭	
768m, b	
713sh	
634w	

References 247, 500

CYCLOSILICATES

505
BENITOITE
$2[BaTiSi_3O_9]$

1040–28vs, b } b doublet
940–30vs, b } b doublet
920sh ? } b doublet
761vs
572w
485s } b doublet
445s } b doublet

References 4, 9, 18 (reflection), *364*

506
DIOPTASE
$6[CuSiO_2(OH)_2]$

1635w
1005–992vs } variable intensity } b
960–52sh } variable intensity } b
938–23sh } variable intensity } b
890–81w, sp, sh } b
780m } b doublet
773m } b doublet
734–30vw
692vw
612sh
576–73s, b
558–43w, sh
518m, b
457m
412m

References 9, 116, 364, 500

507
EUDIALYTE
$4[Na_4(Ca,Fe)_2Zr(OH,Cl)_2Si_6O_{17}]$

1650vw
1055sh } b
1028vs } b triplet } b
977vs } b triplet } b
936s, sh } b triplet } b
910sh } b
840sh } b
742m
700m, b, sh
660m, sh
610vw
580vw
555sh
545m, sh
480s
455m, sh

References 9, 364

508
EUCOLITE
Var. Eudialyte (Nb rich)
$4[Na_4(Ca,Fe)_2Zr(OH,Cl)_2S_{\cdot 6}O_{17}]$

1635w
1510vw
1430vw
1130sh } vb
1055sh } vb
1022sh } vb
977vs } vb
935m, sh } vb
840sh } vb
740m } b
700m, b, sh } b
665b, sh } b
580sh
535m, b, sh
480s
455m, sh

Reference 10

509

CATAPLEIITE

$2[(Na_2Ca)Zr(Si_3O_9) . 2H_2O]$

3500
1650w
1065sh
1025–20vs } b doublet
958–50vs }
745–38m
435m

References 9, 364, 500

510

ELPIDITE

$Na_2Zr(Si_6O_{15}) . 3H_2O$

1650m
1170–45m, sh }
1126–20s, b, sh } vb
1035b, sh }
1012–06vs, b }
808m, sp, sh
785–80m, sp
740vw
708m, sh
647–44m, sh } b doublet
630–25s }
595m, sh
543m
522m, sh
493sh
450sh
433vs

References 10, 50, 364

511

XONOTLITE

$[6CaSiO_3 . 2H_2O]$

3611
1610w
1435w
1210–00s }
1080–70s, sh }
1010–00sh } vb
975–70vs, b }
928–15m, sh }
740vw
710vw
672–70m }
635sh } b
612m, b }
546s
465vs, b
413vw

References 3, 10, 50, 345, 363, 421, 500

512

ALAMOSITE

$12[PbSiO_3]$

1093w }
1025sh }
995m }
948s } b
910sh }
893sh }
846m }

Reference 198

513
WOLLASTONITE
$6[\beta\text{-}(CaSiO_3)]$

ca. 3400w, b
1640?
1105–1087s
1060–56s
[1105–1087s and 1060–56s: b doublet]
1038–30sh
1024–19s
970–64s, sh
930–20sh
905–899vs
[1105–1087s to 905–899vs: b]
687–80m
650–42s
570–66s
515–08m, sh
492–90sh
478–75w, sh
470–65s
455–50s
[470–65s and 455–50s: doublet]

References 7, 9, 50, 78, 385, 467, 500, 585

514
BUSTAMITE
$6[(Mn,Ca)(SiO_3)_2]$

1095–86vs
1062–55m, sh
979–70sh
950–45vs
911–08vs
[950–45vs and 911–08vs: doublet]
[1095–86vs to 911–08vs: b]
872s?
850sh?
803–00sh
780w
725w
670–56w, b
619–15w, b
562–55m, b, sh
530–25sh
517–14w, sh
510–495vs
465–60s, sh
458–50s, b
435sh
415sh
[670–56w, b to 415sh: vb]

References 10, 78, 385, 434, 467

515

FOWLERITE

Var. Rhodonite

10[$(Mn,Zn)SiO_3$]

1630w			610s, b
1095sh	b	doublet	580m
1075sh			565sh
1065s			535sh
1040sh			520inf
955sh	b		495m, sh
950m,			455s
920sh			420sh
900s			
875 sh			
725sw			
670w			
650sh			

Reference 10

516

RHODONITE

10[$(Mn,Ca,Fe)SiO_3$]

3660		
3440–00w, b		
1628		
1090–75sh	b	vb
1064–55vs		
1035–25m, sh		
1003sh	b	
959–50s		
920–15s, sh		
908–02sh		
895–90sh		
876–70sh		
732–23w		
710–695vw		
676–58m, b		
637?		
580–78m	doublet	
563–55m		
537–23sh		
520–15b, sh		
500–495s, sh		
468–55vs		
415sh		

References 3, 4, 78, 385, 434, 452 (poor), *500, 606*, var. Penwithite *3*

517
PYROXMANGITE
16[(Mn,Fe)SiO_3]

1085–80b, sh ⎫
1063–50vs
1026–23sh
1002–998sh
960–48vs, sh ⎬ vb
916–08sh
900–890sh
870sh ⎭
730–20w
695–90w
674
668–61w
645–38sh
577–75m, sh ⎫
563–59sh
532–23sh
514–11m ⎬ b
498–92b, sh
470–58s, b
412vw ⎭

References 10, 78, 385, 452 (poor), *500*

518
PECTOLITE
2[$Ca_2NaSi_3O_8OH$]

3450–00w, b
1610
ca. 1400w
1060 ⎫ vs, ⎫
1040 ⎬ b
1010 ⎭ triplet
998–95sh ⎬ b
974–65sh
928–20vs ⎫ doublet
905–00vs ⎭ ⎭
820–00vb, sh
677–71m–s ⎫ doublet
652–40m–s ⎭
570vw
532–30m ⎫
500sh
488–85m, sh ⎬ b
460–50vs
450–48sh
416sh ⎭

References 9, 50, 51, 385, 500

519
STEVENSITE
$(Mg,Fe)_3Si_4O_{10}(OH)_2$

ca. 3400m, b
2860m, sh
1615w, b
1450w, sh
1370w, b
1110inf
1020vs, vb
870w, b
785w, b
775w, b

Reference 100

520
SCHIZOLITE
var. Pectolite
2[$(Ca,Mn)_2NaSi_3O_8OH$]

1650w, b
1051m, sh ⎫
995sh
960sh ⎬ vb
925vs ⎫ b
890vs ⎭ doublet ⎭
555m, vb ⎫
520sh ⎬ b
505sh
445m, b, sh ⎭

Reference 10

521
BABINGTONITE
$2[Ca_2Fe^{II}Fe^{III}(Si_5O_{14}OH)]$

1660vw			
1275w			
1100m, sh			
1045s	b		
1018sh			vb
1000sh		doublet	
950s			
905m, sh	b		
762w			
705w			
654sh			
645w			
550m, sh			
510m			
495sh			
453sh	vb		
441m, b			
428sh			

Reference 10

522
NEPTUNITE
$8[(Na,K)_2(Fe^{II},Mn)TiSi_4O_{12}]$

1126–18s, sh		
1075–70sh		
1040–38vs, b		
1020?		
990–86sh		
960–54sh	doublet	
935?	?	b
910–01m, sh	doublet	
893?	?	
868–66s, sh	doublet	
856?	?	
837–36m, sh	doublet	
826?	?	
754–52m, sp		
690–84m		
655–50w		
590sh		
560–58m, sh		
530sh	b	
500vs, b		
475sh?		
433m, sh		

References 9, 247

523
BERYL
$2[Be_3Al_2Si_6O_{18}]$

3690s, sp
3590s, sp, sh
3450s, b
1210–00vs
1094–86sh
1035inf
1025–15m, sh
975–55vs, b
} b
810–05s
746–43m
693–80s
655–50w, sh
595–88m, sh

535–24s, vb
495–80m–s
440s ?

Aquamarine
References
17, 130, 325, 364
Emerald *Ref. 405*

References 1, 3, 9, 21, 36, 172, 325 (influence of alkali metal), *347, 364, 548, 585, 617, 631*

524
VOROB'YEVITE
$2[(Be,Cs)_3Al_2Si_6O_{18}]$

1624w
1195sh
1184s
1136sh
1098sh
1060s
1025sh
956vs
} b
813m
794sh

779sh
748m
682m
650sh
597m
524s, b
495s, sh

References 172, 325, 364

525
TOURMALINE
$[Na(Mg,Fe,Mn,Li,Al)_3Al_6(Si_6O_{18})\text{-}(BO_3)_3(OH,F)_4]$?

1450w
1345–30s, sh
1290–80m–s
1185w, sh
1115–10s, b, sh
1030s, b
1000s, b
} b
870w
787–80w–m
750m, b, sh
725–10m, sh

520s, b
475s
425b, sh

References 3, 9, 21, 36, 120, 364, 548, 585, var. Dravite *533* (Xtal-reflection)

526
RUBELLITE/ELBAITE
Formula as Tourmaline

1365–45m, sh
1310–1287s
} b
1190–82sh
1120–04s
1040–35s
1008–998vs, b
} b
900w, sh
790–86m
760w
722–20w

722–20w–m
630b, sh
588m, vb
535–10vs, vb

References 9, 120, 533 (Xtal-reflection)

527
CORDIERITE
$4[(Mg,Fe^{II})_2Al_4Si_5O_{18}]$

1650	674w
1175–70s } b	617w?
1143sh } b	582–77w
1105sh } b	568–65sh
1024sh	487–84s, sh
990sh	448–42sh
965–60vs, b	434s } b doublet
925–10vs, b	415s } b doublet
770–68s	

References 1, 9, 36, 364, 430, 460, 548

528
BAOTITE
$Ba_4(Ti,Nb)_8ClO_{16}(Si_4O_{12})$?

1106vs
962s, sh
926vs
789sh } b
770sh } b
745s, b } b
656s
603s, b
588s
568s
517sh
491s
470

References 247, 500

529
TOBERMORITE
$4[Ca_5Si_6O_{16}(OH)_2 . 4H_2O]$

3500–3450m	585b, sh } vb
1650–15w, b	534m, sh } vb
1445vw	488sh } vb
1210sh	465s } b doublet } vb
1060–50b, sh	450s } b doublet } vb
980vs	440sh } vb
890–65b, sh	414m, sh } vb
745vw	
667–60w, b	

References 10, 345, 421

INOSILICATES

530

ASTROLITE

$(Na,K)_2Fe^{II}(Al,Fe^{III})_2Si_5O_{15} \cdot H_2O$

3610
3440
1640w
1085sh
1035s, b
915m, sh
} b
832vw
805vw
780vw
755vw
700w, b, sh
630sh
525s, sh
473s
430sh

Reference 9

531

CHKALOVITE

$Na_2Be(Si_2O_6)$

1132s, sh
1075sh
1044sh
1008vs
915sh
772m, sp
731m, sp
650m
606sh
585s, b
540m
465m, sh

Reference 172

532

MULLITE

$nr[Al_9Si_{13}O_{19.5}]$?

3356
3000m
2700m
1183–78s, b
1145sh
960b, sh
920–15vs, b
870–47b, sh
750m, b, sh
} vb
620m, b, sh
560–50vs, b
510sh
450–35sh
} vb

References 10, 191, 264, 295, 334, 459, 585

533

CARPHOLITE

$MnAl_2(OH)_4Si_2O_6$

3628s, sp, sh
3590s, sp
3430sh
1086vs, sh
1046m, sh
1010
995
} m, sh
966vs
910m, sh
873m, sh
775sh
740m, sh
710m, b, sh
670m, sh
625sh
585s, b
520sh
477s
427s, sh

Reference 9

534
ENSTATITE
16[$MgSiO_3$]

1180w, sh
1128–04sh
1070–56vs } var.
1020–10s, sh } var.
980–70sh
942–28vs, b
908–02s (sh?)
877–52vs, b
763–58sh
745sh
728–19sh
695–93m, sh
655–647w, b
588–85sh
562–57sh
535–26sh
505m, b, sh
460s, b
410m, sh

References 9, 35, 301, 592, 645 (Xtal-reflection)

535
HYPERSTHENE
16[$(MgFe)SiO_3$]

1630vw, b
1170sh } vb
1110inf } vb
1070vs, b } vb
1028–10m, b, sh } vb
945vs, b } vb
895 } b triplet } vb
872 } sh } vb
855 } vb
760w
725w–m, b
692–90w, b
650w, b
635b, sh
562sh
536m, sh
508vs
456m, b, sh
408?

References 9, 36

536
CLINOENSTATITE (*S*)
8[$MgSiO_3$]

1197sh
1152sh
1102sh
1076vs
1012s, sh
992w, sh
942vs, vb
900s, sh
858s
800w, b
737w } doublet
725w } doublet
684w

References 35, 296

537
AUGITE
8[$(Ca,Mg,Fe^{II},Fe^{III}Al)(SiAl)O_3$]

3425–20w
1630–20w
1068vs, b } vb triplet
970–63vs } vb triplet
918–15m, sh } vb triplet
875–60vs } vb triplet
675–68vw
637–28w
528sh
515–10m, b, sh } b
480–65vs, doublet? } b
410m, sh

References 1, 3, 9, 33, 385, 645 (Xtal-reflection)

538
FASSAITE var. Augite
$8[(Ca,Mg,Fe,Al)(Si,Al,Fe)O_3]$?

1070s	b triplet
1035sh	
970s	
925b, sh	
870s	
790sh	
670sh	
640b, sh	
515m, sh	
475s, b	
440sh?	
412m, sh	

References 9, 643

539
AEGIRITE (INE)
$4[NaFe^{III}(SiO_3)_2]$

ca. 1100b, sh	vs, vb
1060–55sh	
1020–04sh	
965–50s, b	
912–10sh	
897sh	
864–60m, sh	
735–25vw	
643–39m	
562–60s	
545sh	
507–05m, sh	
467–65vs, b	

References 3, 9, 21, 35 (poor), *79, 500*

540
ACMITE
$4[NaFe(SiO_3)_2]$

1048	multiplet b, sh	vb
1012		
954vs		
ca. 860sh		
730vw		
ca. 642m		
568s, sh		
502m, sh		
451vs		

Reference 1

541
DIOPSIDE
$4[CaMgSi_2O_6]$

1120–17sh	vb	677–69m
1092sh		645–35m
1080–70vs		565–64sh?
983sh		520–12s, sh
970–65s		475s, b
930–22s, sh		405m?
870–62s		

References 1, 9, 36, 301, 467, 628, 643

542
HEDENBERGITE
$4[CaFeSi_2O_6]$

3420w
1620w
1090sh ⎫
1070–56vs ⎪
990sh ⎬ vb
970–59sh ⎪
920–11m, sh ⎪
870–60s ⎭
670–63w
635–22w–m
518–09s, b, sh ⎫
500–492sh? ⎬ b
475–54vs, b ⎭

References 1, 9, 79, 119, 385, 467, 500

543
DIALLAGE?
$Ca_7Fe^{II}Mg_{6.5}Fe^{III}{}_{0.5}Al(Al,Si)_{16}O_{48}$

3430
1635w
1080w, sh
1020vs, b
780w ⎫
760w ⎭ b, sh
675m
520inf
460vs, vb

Reference 9

544
JADEITE
$4[NaAl(SiO_3)_2]$

1131sh ⎫
1090–70b, sh ⎪
1065–50vs, b ⎪
995–90m, sh ⎬ vb
940–26s, b, sh ⎪
860–58m, sh ⎪
748–44vw ⎭
718sh?
663–60m, sh
615m, sh
590s, b
535–15b, sh
475–63vs, b
442–29sh

References 1, 3, 35, 84

545
OMPHACITE
$8[(Ca,Mg,Fe^{II},Al)(Si,Al)O_3]$

1075sh ⎫
1044b, sh ⎪
995sh ⎬ vb
966vs, b ⎪
852sh ⎪
742vw ⎭
694w, sh
646w
570m, sh
508sh
465vs, b

Reference 1

546
SPODUMENE
$4[LiAl(SiO_3)_2]$

1139–32b, sh

1095–73vs	var. intensity, positions depend on whether α or β variety
1049–35sh	
1017vs–sh	

α	β
920–17m, sh	896vw, sh
870–62m, sh	765m, vb
642s, b } b doublet	698sh
596s, b } b doublet	560s, vb
542–35b, sh	470–30vs, vb
481–75vs, vb	

References 9, 35, 84

Kunzite Identical spectrum to Spodumene (*Ref. 9*)

547
CHRYSOCOLLA
$CuSiO_3 . 2H_2O$

3410s, vb
1640–30w
1040vs, b
1015sh ?
790vw, b
675w
500s, vb } b
475b, sh } b

References 9, 116, 162, 595

548
RIVERSIDEITE
$Ca_5Si_6O_{16}(OH)_2 . 2H_2O$

3500	760b, sh } vb
3410	675b, sh } vb
1655vw, b	650m, sh } vb
(1460sh) } ?CO_3	630sh } vb
(1405m,) b } ?CO_3	607m, sh } vb
1090sh } vb	580m, b } vb
1040vs } vb	535sh } vb
960sh } vb	498sh } vb
870m, sh } vb	468s } plat } vb
	453s } plat } vb
	430sh } vb

Reference 10

549
INESITE
$[Ca_2Mn_7Si_{10}O_{28}(OH)_2 . 5H_2O]$

3632–20m
3605sh
3430–20s
3250w, sh
1645–40m
1090–87s, sh } b quartet
1055–50s, sh } b quartet
1020sh } b quartet
1000–999s, sh } b quartet
965sh } b quartet
937–35vs } b quartet
910sh } b quartet
773–72vw } vb
727–23vw } vb
670–63m, b } doublet } vb
615–11m, b } doublet } vb
551–49m, b, sh } vb
496–95m, sh } vb
475–65sh } vb
462–50vs } vb
415sh } vb

References 10, 385, 434

550

RACEWINITE

$H_2CaAl_{10}Si_{12}O_{36} . 19H_2O$

(Mixture?)

(approx.)

ca. 1100sh	b
1020vs, b	b
910m, sh	b
ca. 700m, b	

Reference 227

551

NEOTOCITE

$Mn^{II}{}_2Fe^{III}{}_2(Si_4O_{13}) . 6H_2O$?

(approx.)

ca. 3450m, vb
ca. 1610w
1040sh
ca. 1000vs, vb

Reference 92 (poor)

552

ANTHOPHYLLITE

$4[(MgFe)_7(OH)_2(Si_4O_{11})_2]$

1640vw		693w	
1125sh	b	660w	
1100s, sh	b	534m, sh	b
1055sh	b	500s	b
1008sh	b	466sh	b
985vs	b	442m, sh	b
908m, sh	b	413sh	b
782w, sp			
756vw			
738vw			
708m, sh			

References 9, 18 (reflection), *51, 508, 529, 623*

553

AMOSITE

$4[(Fe^{II},Mg,Al)_7(Si,Al)_8O_{22}(OH)_2]$

3630		655sh
3615		640w, b
1129s, sh	b	530sh
1080m, sh	b	500s, b
1000vs, b	b	485sh
975sh	b	440sh
890m, sh	b	428vs, sh
825vw		
776w		
728sh		
700w, b		

References 10, 35 (poor), *508, 529*

554

CUMMINGTONITE

$2[(MgFe)_7Si_8O_{22}(OH)_2]$

1126m, sh	vb	519sh	b
1085s, sh	vb	492s	b
1000sh	vb	455sh	b
978vs	vb	433s	b
900m, sh	vb		
778w			
740sh			
690w			
648w			

References 1, 623

555

HORNBLENDE

$2[(Na,K)_{0-1}Ca_2(Mg,Fe^{II},Fe^{III},Al)_5$-
$(Si_{6-7}Al_{2-1}O_{22})(OH,F)_2]$

1100–1080s, sh		690–80w	
1060–40sh		640–30w, sh	
986–80s	b doublet	510–05s, sh	b
955–30s	b doublet	465–62vs	b
919–08sh			
755–40w–m			

References 1, 9, 33, 35, 276, 623

556

RICHTERITE

$2[(Na,K)_2(Mg,Mn,Ca)_5Si_8O_{22}$-
$(OH)_{22}]$

(approx.)
ca. 760m
ca. 690w
ca. 650w
ca. 500vs, b

Reference 452 (poor)

557

HASTINGSITE

$2[Ca_2Na(Fe,MgAl)_5(OH)_2$-
$(Al,Si)_8O_{22}]$

(approx.)
ca. 1100sh
1000vs
750w
700sh
530sh
470vs

Reference 276

558

AÇTINOLITE

$2[Ca_2(Mg,Fe)_5Si_8O_{22}(OH)_2]$

3668
3654
3540
3430
1105–1094s, sh
1060–52m, sh
1037sh
1012–995vs } b
955–48vs } doublet
925–14m, sh
760–55w–m
727?
686–82w
660–55sh
540sh
510–07m, sh
476 } vs, b
466–62 } doublet?
453–45sh

References 1, 9, 33, 51, 119, 500, 508, 529, 623

559

TREMOLITE

$2[Ca_2Mg_5Si_8O_{22}(OH)_2]$

1110–05vs }
1075–60vs }
1048sh? }
1020–10sh } vb
1000–995vs }
955–50vs }
925–20s, sh }
758–55m, sp
730–20w, sh
690–82m } doublet?
670–62m, sh }
646–45w
550–45sh
532sh
515–10sh
470–66vs, b
455–51sh
425sh

References 9, 33, 36, 51, 301, 500, 508, 529, 623

560

NEPHRITE

$2[Ca_2(Mg,Fe)_5(OH)_2Si_8O_{22}]$

3670
3655
3430
1630vw
1107vs }
1062vs }
1020sh? } b
1002vs } quintet
955vs }
925s, sh }

758m, sp
725sh
680m
665m, sh
646m, sh
540sh
510s, sh
465vs
450sh?
420m, sh

Reference 9

561

GLAUCOPHANE

$2[Na(Mg,Fe^{II})_3Al_2Si_8O_{22}(OH)_2]$

1145sh }
1102s, sh }
1045s, sh } vb
996sh }
978vs }
880sh }
790m
730sh
695w } b
673w } doublet

ca. 650sh
555m, sh }
515sh } b
480s }
440sh }

References 1, 3, 9

562

RIEBECKITE

$2[Na_2Fe^{II}_3,Fe^{III}_2Si_8O_{22}(OH)_2]$

3612
3430
1630vw?
1124sh }
1078–70m, sh }
1030sh } vb
994sh }
980–64vs }
900–885sh }
775w } b
754–52w } doublet

710–694vw
670sh
650–36w
537m, sh
492–80sh
452–50vs

References 1, 3, 9

563

ARFVEDSONITE

$2[(Na,Ca)_{2.5}(Fe^{II},Fe^{III},Mg)_5Si_8O_{22}$-
$(OH)_2]$

3440
1630vw
1126m, sh }
1060sh }
960vs } b
890sh? }
865sh }
742w
660sh
642w

490sh }
450s } b
420sh }

References 9, 623

564
CROCIDOLITE
$2[Na_2Fe^{II}{}_3Fe^{III}{}_2Si_8O_{22}(OH)_2]$

3647m, sh
3631s, sp
3620s, sp
1144m, sp, sh ⎫
1106s ⎪
1048sh ⎪
991vs ⎬ vb
978sh ⎪
895m, sh ⎪
881sh ⎭
777m
727vw
693w, b
656vw
623w, b

References 3, 362, 508, 529

565
HOLMQUISITE
$Li_2(Mg,Fe^{II})_3(Al,Fe^{III})_2(Si_8O_{22})$-$(OH)_2$?

(very approx.)
ca. 1100sh ⎫
1030vs ⎬ b
880sh ⎭
800w
740w, b
600sh
580sh
560sh
500vs, doublet?

Reference 276

566
CHEVKINITE
$(Fe,Ca)(Ce,La,Al)_2(Si,Ti)_3O_{10}$

(approx.)
1630m
1400m
1120sh
980vs, b
550sh ⎫
520m ⎬ b
450m ⎭

References 226, 627

567
SEPIOLITE
$Mg_3Si_4O_{11} . 5H_2O$

3690–40
3570
3440–10
1640–38w
1200–1195sh
1120sh
1072–63sh
1030–25vs
1008–975m, sh
890sh
872–65sh
798–82b, sh
685b, sh
640m
535sh ⎫
475s ⎬ b
443m, sh ⎭

References 9, 34 (poor), *35, 209, 227, 368, 484*

568
EPIDIDYMITE
$8[NaBeSi_3O_7(OH)]$

3520w, sp } doublet
3470w, sp }
1665w
1148s, sp }
1119sh }
1085sh } b
1065sh }
1013vs }
884sh
867w
795s, sp
750m, sp
633w
602w
505sh

Reference 172

569
EUDIDYMITE
$8[NaBeSi_3O_7(OH)]$

3520w } doublet
3470w }
1655w
1155–41s }
1122–16sh }
1096sh } b
1048–45sh }
1021–15vs }
984sp, sh }
879–75w
795–92s
765sh
738–34w
728sh
701m
698–77sh } var.
638–25w }
604–599sh
588–73m
507–00sh
488–74sh
460s, sh
422vs
407sh

References 10, 172

570
ZOISITE
$4[Ca_2Al_3Si_3O_{12}(OH)]$

3140		595sh	
1115vs, b	b	575vs, b	
1040sh	b	512s, sh	
975sh	b	470s	b
950vs	b	443s	b
900m, sh	b	410s ?	
865sh	b		
780sh			
755m, sh			
715sh			
695b, sh			
655s, sh			
620m, sh			

References 9, 52, 631

571
EPIDOTE
$2[Ca_2(Al,Fe)_3Si_3O_{12}OH]$

3360
1115–02s
1080–68sh
1040–25m, sh
950–47vs
885s, sh
860sh
838sh
720m, sh
650s, b
570b, sh
520s, b
455m, b, sh

References 9, 250, 548, 627

572
PISTACITE
$2[Ca_2(Al,Fe)_3(OH)Si_3O_{12}]$

3340–00m, sp		661
1610vw		575
1102–1098s	vb	527
1068–60m, sh	vb	404
1025–990s	vb	
951–47vs	vb	
901–885s	vb	
862		
823		

References 52, 250

573
PIE(D)MONTITE
$2[Ca_2(Al,Fe,Mn^{III})_3Si_3O_{12}(OH)]$

1110s	b triplet sh	vb	710sh
1070s	b triplet sh	vb	650m, vb
1045sh	b triplet sh	vb	570b, sh
1030s	b triplet sh	vb	522sh
990sh		vb	516s, b
960s, b		vb	502sh
890s		vb	468s, b
840m, sh		vb	
800sh		vb	

Reference 10

574
ALLANITE
$(Ca,Fe^{II})_2(Al,Ce,Fe)_3Si_3O_{12}OH$

(approx.)
ca. 1450w
ca.1050vs
910vs
850sh
670sh
620m
550m, sh
500s

Reference 226

575
ORTHITE
$(Ca,Ce,La,Na)_2(Al,Fe,Be,Mg,Mn)_3$-
$Si_3O_{12}(OH)$

1045m, b, sh }
950s, vb } vb
865m, sh }
640vb, sh }
580vb, sh } vb
510s, vb }
460b, sh }

Reference 10

		Reference
Catophorite	nr $2[(Ca,Na,K)_3(Mg,Fe^{II},Fe^{III},Al)_5(SiAl)_8O_{22}(OH)_2]$	*623*
Dannemorite.	$2[(Fe,Mn,Mg)_7Si_8O_{22}(OH)_2]$	*623*

PHYLLOSILICATES

576
GILLESPITE
$4[BaFe(Si_4O_{10})]$

Band	Grouping
1230b, sh	vb
1165m, sh	vb
1080sh	vb
1052s, sh	vb
1000s	vb
803vw	
755vw	
662m	
608w, b	
558b, sh	
512sh	
465s	b doublet
430s	b doublet
412sh	

Reference 10

577
EUCLASE
$4[AlBeSiO_4(OH)]$

Band	Grouping
3580	
3440	
1640vw	
1072vs, sh	
1028sh	
980b, sh	
915vs	
875sh	
796s	b doublet
776s	b doublet
742sh	
682s	b doublet
665s	b doublet
612sh	
595s	
570m, sh	
540s	
480m	
450s, sh	
427s	

Reference 9

578
SANBORNITE
$BaSi_2O_5$

Band	Grouping
1138sh	vb
1089s, sp	vb
1038sh	vb
998s	b doublet; vb
978s	b doublet; vb
800vw	
759w	
694w	
621vw	
540m	b doublet?
534m	b doublet?
483	
468s	b doublet?
460s	b doublet?

References 3, 10, 500

579
OKENITE
$CaSi_2O_4(OH)_2 . H_2O$

Band	Grouping
3580	
3460	
1405vw	
1150sh	vb
1110sh	vb
1092sh	vb
1073–65s, b	vb
1046–38sh	vb
1017–04m, sh	vb
985sh	vb
938–30sh	vb
910s	vb
773vw	
744vw	
725vw	
667–62m	b doublet
650–44m	b doublet
604sh	
545–30m	
515sh	
493sh	
475sh	
453–50s, b	

References 10, 421, 500

580

NARSARSUKITE

$4[Na_2(Ti,Fe)(O,OH,F)Si_4O_{10}]$

1080b, sh	650w
1040sh	570m
1020s	547m, sh
960s, b	507m, sh
940sh	455s, b
910sh	430sh
863m, sh	
738m, sh	

(1080–738: vb; 570–547: b; 650–430: vb)

References 10, 500

581

FRIEDELITE

$6[(Mn,Fe)_8(OH,Cl)_{10}Si_6O_{15}]$

3600	728s, sh
1152s, sh	694m, sh
1124sh	653m, sh
1050vs, b	632m
1030sh	546w
970m, sh	500m
	450s, b

(1152–970: b; 728–632: b)

Reference 10

582

PYROSMALITE

$2[(Mn,Fe^{II})_8(OH,Cl)_{10}Si_6O_{15}]$

3580	540m, sh
3550	512m
1142s, sh	436s
1135sh	413sh
1055vs	
1015m, sh	
960sh	
765sh	
716sh	
670m	
640b, sh	

(1142–960: vb; 765–640: b)

Reference 10

583

PREHNITE

$2[Ca_2Al_2(OH)_2(Si_3O_{10})]$

3485	640vw
3475	530s
1630vw	500b, plat
1090	475m, sh
1075	426s
995vs	
940s, sh	
868w	
815w	
745m	
675w, b, sh	

(1090–1075: s, b, sh)

Reference 9

584
PYROPHYLLITE
$4[Al_2Si_4O_{10}(OH)_2]$

3668
3640
3440
1125–14s, sh
1070vs } b
1058–42vs } doublet
951–39s
856–54w }
837–36w } triplet
819–01w }
781–79vw
738vw
627–24sh
580–78m, sh
542–35s
522–17m, sh
485–80s
465–62sh
421–19m, sh

References 9, 35 (poor), *40, 49, 53, 88, 100, 295, 301, 368, 370, 398, 469, 479, 502, 508*

585
TALC
$4[Mg_3(Si_4O_{10})(OH)_2]$

3685
3676
3660
1640–20w
1050–40sh
1030–18vs, b
1012–05vs
783–70w
740
700–690sh
675–68s, sp
540–33s, sh
500inf }
475–57m, sh } doublet? } b
455–45vs } doublet? } b
444–40inf }
428–22m, sh }

References 1, 9, 33, 40, 49, 88, 100, 119, 201, 295, 301, 327, 368, 370, 398, 469, 508

586
WILLEMSEITE
$4[(Ni,Mg)_3(Si_4O_{10})(OH)_2]$

3676vw
3662w
3645s
3624s
1025vs, b
705s } b
668s } doublet
465vs } b
452vs } doublet
413s, sh
406sh
383s, sp, sh
358m, sh

References 398, 456

587
MINNESOTAITE
$8[(Fe^{II},Mg)_3Si_4O_{10}(OH)_2]$

(approx.)
ca. 3500w, sp
1180sh } b
ca. 1010vs, b } b
830w
790w
785sh
660m
605sh
580sh
470s } b
460s } doublet
417sh

Reference 95

588
MUSCOVITE
$4[KAl_3Si_3O_{10}(OH)_{12}]$

3730
3630
3448–30
1075–60sh
1030–21s, b
1000–991sh
940–24w, b, sh
831–27w
810–00vw
755–50w
692–87vw, b
535–00s } b doublet
480–70s }
435–15b, sh

References 1, 3, 9, 21, 33, 40, 88, 188, 195, 201, 295, 301, 306, 347 (Xtals), *368, 370, 480, 508*

589
PHENGITE/
PICROPHENGITE
$4[K_2(MgAl)_2(Al,Si)_4O_{10}(OH)_2]$

1175sh }
1085sh }
1050vs } vb
980sh }
935sh }
815w } b
780w }
700vw
ca. 595sh
525m, sh
465vs

Reference 88

590
PARAGONITE
$4[NaAl_3Si_3O_{10}(OH)_2]$

ca. 3640m
1082sh
1038 } b doublet
1015 }
994vs, b
930sh
875sh
834sh
808–03w
755–50w–m
695sh
650sh
625–15sh
590b, sh
545–35vs } b doublet
485–78vs }
417s, b, sh

References 1, 9

591
LEPIDOLITE
$4[K(Li,Al)_3(Si,Al)_4O_{10}(OH,F)_2]$

3625
3440
1129s }
1100–1084b, sh } b
1026–22vs }
997–85vs }
968–66m, sh
805–790w
760–54w
688vw
583m, sh
553sh
535–29s, b, sh }
482–75s } b
445–35m, sh }

Position of absorption bands depends on the aluminium content.

References 1, 9, 301, 370

592

ZINNWALDITE

$2[K_2(Li,Fe^{II},Al)_6(F,OH)_4$-
$(Si,Al)_8O_{20}]$

3571–40
3440
1802vw
1640vw
1428w
1100b, sh? }
1040–28vs, b } b
1010–00inf }
780sh }
740w } b
702w }
530m, b, sh
435s
410m, sh

References 9, 64 (poor), *301*

593

BIOTITE

$2[K_2(Mg,Fe^{II},Al,Fe^{III})_{4-6}(Si,Al)_8$-
$O_{20}(OH)_4]$

ca. 3700w, b
3448
1620–13w
1067–50sh
1023–988s, b
975–70sh
ca. 920sh
780–60sh } w, vb multiplet?
700–692m }
612b, sh
462–49s, b

References 1, 9, 33, 40, 88, 245, 295, 301, 370, 406 (composition correlation), *444, 445*

594

PHLOGOPITE

$4[KMg_3(Al,Si_3)O_{10}(OH,F)_2]$

3650
3450–40m, b
2860m, sh
1630–15w, b
1430w, vb
1075inf
1010–990vs, vb
980sh
900sh
815–05s, b
725sh
695–90s
675m
615–10vb, sh
466s, b

References 1, 9, 88, 100, 301, 370, 445

595

GLAUCONITE

$2[(K,Na,Ca)_{1.2-2}(Fe^{III},Al,Fe^{II},Mg)_4$-
$(Si_{7-7.6}Al_{1-0.4}O_{20}(OH)_4 . nH_2O]$

3560
3540
3410
1630vw
1270sh
1110sh
1030–00vs, b
810–05vw
750vw
680–75w
610sh }
575b, sh }
500m, sh } vb
480–70s }
450–40sh }

References 1, 9, 201, 226, 585

596
CHRYSOTILE
$2[Mg_3Si_2O_5(OH)_4]$

3680m, sp
3350sh
1630w, b
1085–75s } b triplet
1020–10s, sh } b triplet
972–55vs, b } b triplet
605b, s, b } b
536sh } b
465b, sh
427vs
412sh

References 20, 49, 62, 88, 99, 301, 398 (Ni–Chrysotile), *508*

597
GARNIERITE
$(Ni,Mg)_3Si_2O_5(OH)_4$

3685
3672
3430
1635vw
1040sh
1020s
672m
605w
536w } b
500w, sp } b
465s } doublet } b
453s } doublet } b
427m, sh } b

References 9, 116

598
GREENALITE
$Fe^{II}_9Fe^{III}_2Si_8O_{28} . 8H_2O$?

3500w, sp
1180sh } b
1040vs, b } b
835vw
785w
770sh
670m } b
610sh } b
580sh } b
470s, b } b
450sh } b
410 } b

Reference 95

599
Al–CHLORITE?

3620s } b doublet
3520s } b doublet
3340m, b
1640vw, b
1010vs, b } b
940sh } b
825vw
692m, vb
555sh } vb
528s, b } vb
475s, vb } vb

Reference 502

600

ROTTISITE

Var. Antigorite

$16[Mg_3Si_2O_5(OH)_4]$?

3630
3440
1635w
1028vs, b
700sh } b
675m }
570s, b
432sh

Reference 9

601

PENNINITE

$2[(Mg,Fe^{II}Al)_6(Si,Al)_4O_{10}(OH)_8]$

3610–3590m
3470–30m
1635w
1080sh
1050–40m, sh
1000–982s } b
970–58s }
821–20w
755sh
660–56m, b
525sh
460–45vs, b

References 9, 168, 221 (poor)

602

SHERIDANITE

Var. Clinochlore

$2[(Mg,Al)_6(Si,Al)_4O_{10}(OH)_8]$

3660vw
3570m
3425m
1630w
1080vw
989s
825w
766m
674m

Reference 168

603

DIABANTITE

$2[(Mg,Fe^{II},Al)_6(OH)_8(Si,Al)_4O_{10}]$

(contains quartz)

3610m
3470m
992s
945sh
647m quartz absorption omitted

Reference 168

604

CORUNDOPHILITE

$2[(Mg,Fe^{II},Al)_6(Si,Al)_4O_{10}(OH)_8]$

3635vw
3565m
3445m
1630w
983s
816w
761m
661m

Reference 168

605

CLINOCHLORE

$2[(Mg,Fe^{II},Al)_6(Si,Al)_4O_{10}(OH)_8]$

3620–00s	656–55s, b
3460–50s	525b, sh ⎫
1640–35w	460vs ⎬ b
1085–80sh	445sh ⎪
1058–42m, sh	415sh ⎭
1002–984vs ⎫ b	
963–59vs ⎭ doublet	
828–20w	

References 9, *49*, *127* (poor), *168*, *195* (heating)

606

RIPIDOLITE/
PROCHLORITE

$2[(Mg,Fe^{II},Al)_6(Si,Al)_4O_{10}(OH)_8]$

3635	550sh ⎫
3560m ⎫ b	460sh ⎬ b
3425–20m ⎭	435s ⎭
1640–35w	
1016–990vs	
820sh	
760–46w	
667–55w–m, b	

References 9, *168*, *469*

607

THURINGITE

$2[(Fe^{II},Fe^{III},Mg,Al)_6(Si,Al)_4O_{10}$-
$(O,OH)_8]$

3635
3560–50m ⎫ b
3430–3390m, b ⎭
1640–25vw, b ⎫
990–79vs, b ⎪
808–794m ⎫ b ⎪
765–52m ⎭ doublet ⎬ vb
670–61m, b ⎪
620m, b ⎭
550sh ⎫
490sh ⎪
460vs ⎬ b
435s, b ⎭

References 116, *168*, *469*

608

AMESITE

$4[(Mg,Fe^{II})_4Al_4Si_2O_{10}(OH)_8]$

3595m
3560m
3425m
985s
929m
825m
777m
697m
675m

References 62, 168

609

LIZARDITE

$Mg_3Si_2O_5(OH)_4$

3650–35m, sp
3390m, sh
1610vw
1093–70s, b
1000–970sh
952–35vs, b
874w, sh

References 99, 221 (poor)

610

LEUCHTENBERGITE

$2[(Mg,Al)_6(SiAl)_4O_{10}(OH)_8]$

3635vw
3574–70s } b
3450–12sh }
1640–35w
1150sh }
1090–80sh }
1036sh } b
995–91vs, b }
962sh }
887
829–12w }
760–57m } b
661–58s, vb }
548sh }
522sh } b
480s, vb }

References 168, 469, 502

611

CHLORITE (Group name)

$(Mg,Fe,Al)_6(OH)_8(Si,Al)_4O_{10}$

3660w
3560m
3400m
1640w
1081vw
1042w
992–81s
953w
812w
758m
612m

References 168, 585

612
PICROLITE
$Mg_3Si_2O_5(OH)_4$

3620m, sp	646s, sh
3300w, b	610s
1600vw, b	568s
1196w, b	452vs, b
1117sh	
1080s	
985vs	
953sh	
873vw, b	
775vw, b	
705vw, b	

References 88, 99

613
KAEMMERERITE
$Mg_5(Al,Cr)_2Si_3O_{10}(OH)_8$

3660–00w	525m, sh } vb
3585–50m	464vs, b } vb
3460–50m	445sh } vb
1640–35w	413s } vb
1175sh? } b	
1090–80sh } b	
1050–42b, sh } b	
1005–996vs } b	
963–62m, sh } b	
870w	
822w	
657–55s, vb	

References 10, 168

614
CHRYSOPRASERDE/
SCHUCHARDITE
$(Ni,Mg,Al)_6(OH)_8(Al,Si)Si_3O_{10}$

3675	675w, b
3545	600vw
3420	535inf
1635 } b	598sh
1170sh } b	468vs } b doublet
1070sh? } b	455vs } b doublet
1025vs } b	
880sh } b	
830vw } b	

References 9, 116

615
NEPOUITE
$(Ni,Mg)_3Si_2O_5(OH)_4$

(approx.)
1080sh } b
1010b, sh } b
955vs } b

Reference 221

616
CRONSTEDITE
$6[Fe_2^{II}Fe_2^{III}SiO_5(OH)_4]$

3500
3300
1015sh
940vs, b
745sh
675w, b
570vw
445m, vb
412sh?

References 10, 62

617
CHAMOSITE (impure)
$2[(Fe^{II},Fe^{III},Al,Mg)_6(Si,Al)_4O_{10}$-
$(O,OH)_8]$

3550–40m } b
3410–3380m } b
1080b, sh
990s, b
670w, b
618w, b
540sh } b
462s, b } b
430s, sh } b

References 9, 116, 469

618
ANTIGORITE
$16[Mg_3Si_2O_5(OH)_4]$

3680–65m, sp	660sh
3570	620–15s
3425	585–70var, sh
1630w, b	450–42vs, b
1204–00w, b, sh	410sh
1130–15vw, b, sh	
1085–75s, sh	
990–85vs, b	
960–55sh	
773–70vw	
710vw	

References 9, 18 (reflection), *62, 88, 99, 168, 485, 508*

619
STILPNOMELANE
$K(Fe^{II}Fe^{III}Mg,Al)_3(OH)_2Si_4O_{10}$.
$n(H_2O)$

3580
3430
1640vw
1430w?
1200sh } b
1100sh } b
1050vs } b
880m, sh } b
660w, b
440s, b
415sh?

References 3, 10,

620
XANTHOPHYLLITE
var. of Seybertite
$2[Ca(Mg,Al)_{3-2}(OH)_2(Al,Si)_4O_{10}]$

940vs, b } b doublet
905sh } b doublet
805vs, b } b doublet
660vs, b
560b, sh } b
510sh } b
490sh } b
450vs, b } b
425b, sh } b

Reference 10

621
SEYBERTITE
$Ca(Mg,Al)_3(Al,Si)_4(O,OH)_{12}$

(approx.)
3450s, b
ca. 1090b, sh } vb
987vs, b } vb
ca. 935b, sh } vb
890b, sh } vb
830sh } vb
800m, vb } vb

Reference 301

622
MARGARITE
$CaAl_4Si_2O_{10}(OH)_2$

3636m
3624sh
3400m, b
1060sh } b
1025–00vs, b } b
940–26vs } b
900inf } b
860vw } b
830vw } b
812vw } b
800vw } b
750w
724sh
697–95w–m, b
608s, sh
541–40s } b doublet
495–85s } b doublet
415m, b, sh

References 9, 301, 370

623
CHLORITOID
$4[(Mg,Fe^{II})_2Al_4Si_2O_{10}(OH)_4]$

3450s
3340sh
2980m, b
1650vw
1105m, sh
962m, sh
908s
870sh
805m, sh
750s
672m, b, sh
612m } b
590sh } b
552m, sh
518sh
450vs, b

Reference 9

624

GYROLITE

$48[Ca_2Si_3O_7(OH)_2 . H_2O]$

(*N* and *S*)

ca. 3400m
2940w, sh
1665w, b
1490sh
1440m
1125sh }
1085m, sh }
1030–1010vs, b } b
980–70sh }
950sh }
890sh }
795m
770w

References 90, 421

625

TRUSCOTTITE

$6[Ca_2Si_4O_9(OH)_2]$

3850m, sp	695w, b
ca. 3300w, b	500b, sh } b
1235sh } vb	475sh } b
1150b, sh } vb	470s } b
1065sh } vb	455sh } b
1055vs } vb	400m
1040sh } vb	ca. 380sh
1000m, sh } vb	
875sh } vb	
800m } b	
780sh } b	

References 249, 421

626

REYERITE (*N* and *S*)

$nr48[Ca_2Si_3O_7(OH)_2 . H_2O]$

3700m, sp	935–34sh
3400–3350sh	875sh
1615vw	795–90m
1430m, b	775sh
1165sh	730m, b
1135sh	670w, b
1085sh	
1040vs	
1020sh	
1000sh	
980–75sp, sh	

References 90 (Xtals), *248*

627

HISINGERITE

$Fe^{III}{}_2Si_2O_5(OH)_4 . 2H_2O$?

(approx.)
ca. 3300s, b
ca. 1610w, b
ca. 1010vs, vb

Reference 92

628

CELADONITE

$K(Mg,Fe^{II})(Fe^{III},Al)Si_4O_{10}(OH)_2$

	3550m, sp	
ca.	3450sh	
	1620w	
	1115–05sh	b
	1075m, sh	
	972–70vs	
	845vw	
	800–798m	
	750–46vw	
	681s	
	494–85s, sh	b triplet
	460–57s	
	442–38s, sh	
	427sh	

References 1, (Fe–Celadonite) *10*, *258*, *370*

CLAYS (KANDITE GROUP)

629
KAOLINITE
$2[Al_2Si_2O_5(OH)_4]$

3696s, sp		605w, sh
3670–56m, sh		542–35s
3645w		475–68s
3630–24s, sp		435–28m, sh
1117–05s, sh		347w
1035–30s	b doublet	277w
1019–05s		202w, sh
940–35m, sh		190w
918–09s		
800–784w		
760–48w		
700–686m		

References 1, 4, 9, 20, 33, 45, 49, 53, 58, 62, 88, 133, 140, 201, 209, 266 (far-IR), *279* (Xtals), *295, 370, 398, 438, 469, 502, 514*

630
NACRITE
$12[Al_2Si_2O_5(OH)_4]$

3710–3696sp, sh	794m
3658–54s, sp	754m, sh
3627	700–697m, b, sh
3450	605–03sh
1120s, sh	540–30vs, b
1038–34vs	472–70vs
1002vs	430–25m
956vw, sh	415sh
936–32m, sh	
914s	

References 9, 62, 127, 202, 370, 469

631
ANAUXITE
$Al_2Si_3O_7(OH)_4$

3695			792w
3670			755w
3652			695m, b
3618			536s, b
1116	s, sh doublet	vb	468s
1100			430s, sh
1032vs	b doublet		415sh
1010vs			
938sh			
913s			

References 10, 333

632
DICKITE
$4[Al_2Si_2O_5(OH)_4]$

ca. 3700m, sp	
3660s, sh	
3620s, sp	
1620w	
1130–20s	vb
1110–00sh	
1053–36vs, b	
1005–995vs	
945–35m, sh	
914s	
800–792w, b	
752–46w, b	
700–676m, b	
608sh	b
ca. 540vs, b	
475–70s, b	
429s	
415sh	

References 10, 49, 62, 127, 153, 469, 493

633

HALLOYSITE

$2[Al_2Si_2O_5(OH)_4 . 2H_2O]$

3702–3696sp, sh
3629–24s, sp } b
ca. 3414s, vb
3250?
1650–40w
1112–1093sh
1040–35vs
1020–05sh } b
918–10s, sh
830b, sh
797–90m, sh
758–50m, sh } b
723sh?
695–90m, b, sh
ca. 660m, b
545–50s, vb
474–50s
436m

References 9, 53, 58, 62, 127, 202, 209, 295, 438, 469, 585, 496 (heating effects)

634

METAHALLOYSITE

$2[Al_2Si_2O_5(OH)_4]$

3696m, sp
3630–20m, sp
3570sh
3410b, sh
1122m, sh
1045vs, b
1020sh
915s
790w, b
750w, b
680m, sh
650sh
550s
480s
450m, sh
420sh

References 88, 127, 469

635

WOLCHONSKOITE

$Cr_{0.35}(Cr,Fe^{III},Al,Mg)(OH)_2(Al,Si)_4$-$O_{10}(Mg,Ca)(H_2O)_4$?

3420
1640w
1025vs, b
860sh
685b, sh
625sh
525s } b
462s } doublet } vb
415sh?

Reference 10

636

RECTORITE

$(Na,Ca,K)H_3O(Al)_4(Al,Si)_8O_{20}$-$(OH)_4 . 3H_2O$

1121sh
1048vs } b
1020vs } doublet
989sh
916sh
875w, sh
822w } b
808w
747sh
731w } b
705sh
641vw
567sh
515sh
480s } b
431m, sh
411m, sh

Reference 370

CLAYS (MONTMORILLONITE OR SMECTITE GROUP)

637

MONTMORILLONITE

$(\frac{1}{2}Ca,Na)_{0.7}(Al,Mg,Fe)_4(Si,Al)_8$-$O_{20}\ (OH)_4 \,.\, nH_2O$

3644–20s	630–25m, b, sh
3428–3390s	540m, sh
1640–35m	470s
1130–1090sh	430sh
1040–15vs, b	278vw
920–09m, sh	194vw
885–80vw	
850–35w	
800–795w	

References 9, 33, 49, 88, 91, 112, 140, 162, 201, 266 (far-IR), *269, 295, 370, 438, 490, 502, 504*

638

HECTORITE

$Na_{0.67}(Mg,Li)Si_8O_{20}(OH,F)_4$

3670–10 doublet?	
3450–00 var.	
1650–30w	
1430m	
1075–73s, sh	} b
1011–05vs, b	} b
873–53w, sh	} b
820–780sh	} b
712–696m	} b
655m	} b
533–30sh	
470–65vs, b	

References 4, 91, 100, 127, 301, 327, 370, 484

639

CHLOROPAL/NONTRONITE

$(Fe,Al)_9(Si,Al)_{16}O_{40}(OH)_8 \,.\, nH_2O$

3636–20s	686m, sh	} vb
3425–00s	600sh	} vb
1640–35w	497–90s	} vb
1120–1092sh	460–55sh	} vb
1050–20s, b	440–35s	} vb
856–50w, sh		
824–20w–m		
781–60w, sh		

References 49, 91, 99, 112, 127, 162, 370, 398, 585

640

BEIDELLITE

$R_{0.33}Al_2(Al,Si)_4O_{10}(OH)_2 \,.\, nH_2O$

(R = Na, K, Mg or Ca)

3640m, sp	776m
3450m, sh	729vw
1630w	693vw
1084sh	624vw
1041–30vs	566m, sh
1022sh	537s
942m, sh	511sh
877w, sh	479s
819m	426s

References 88, 127, 370, 398

641

STEVENSITE/GHASSOULITE

$\frac{1}{2}(Ca,Mg)_{2x}Mg_{3-x}Si_4O_{10}(OH)_2$

(approx.)
3450s
1640w
1410w
1020 centre vb peak

Reference 100

642

SAPONITE

$(\frac{1}{2}Ca,Na)_{0.67}(Mg,Fe)_6(Si,Al)_8O_{20}$-$(OH)_9 . 4H_2O$

3697w	655m
3670w	540–34m
ca. 3400s, b	490inf
2940m, sh	470–64vs } b doublet
ca. 1620m	450vs } b doublet
1430w, vb	420w
1056vs	
1005–00vs, b	
809–00vw	
784inf	
737sh?	
672m	

References 88, 90, 100, 162, 327, 370, 585

OTHER CLAY MINERALS

643

APOPHYLLITE

$2[KCa_4Si_8O_{20}(F,OH) . 8H_2O]$

3560s
3330b, sh
3040m, vb
1695w
1126m, sh
1095m, sh
1050–47sh
1016vs
790sh
764–62m
630–24sh
602m
553–40m, b, sh } doublet?
536? }
503s
478–75s, sh
445sh
415m

References 9, 50, 301 (poor), *347* (Xtals), *500*

644

PALYGORSKITE

$2[(Mg,Al)_5(Si,Al)_8O_{20}(OH)_2 . 8H_2O]$

3612–00
3575–60
3400
3300
3200
3130
1660–20s
1200–1190sh }
1125–17sh }
1110–1090sh }
1035–25vs, b } vb
1015–975vs }
948? }
915–10sh }
880sh }
640b sh
580sh
512s } b doublet
485s }
444b, sh
415sh

References 10, 35, 40, 209, 295, 368, 487, 585

645

ILLITE

$[(H_3O,K)_4Al_8(Si,Al)_{16}O_{40}(OH)_8]$

3630s
ca. 3400w, sh
1640w
1150sh
1080–72sh
1025–10vs, vb
927–20b, sh
870sh
824–22w, vb
804–797w
777–53w, b
709vw, b
620sh?
530–25vw, vb
470–65vw, vb
435–30b, sh
415vw?
397sh }
370sh }
264sh } b
194vw }
164vw }

References 10, 53, 133, 266 (far-IR), *295, 438, 469, 487, 502* (two types), *585*

646

VERMICULITE

$2[(Mg,Fe^{II},Fe^{III},Al)_{6-7}(Si,Al)_8O_{20}\text{-}(OH)_4 . 8H_2O]$

3522–00m, b
3390?
1640–18w, b
1380w
1017–00vs, b
ca. 660w, vb
ca. 455s, vb

References 153, 269, 295, 368, 469, 508

TECTOSILICATES

647
LEUCITE
16[$KAlSi_2O_6$]

1118–10sh	426w
1038–30vs, vb	479sh
951inf	446–36s, b
830vw	415sh
770–67m, b	
735–22m, b	
642–29w	
613w, sh	
549–45w	

References 84, 497, 585, 582

648
POLLUCITE
16[$(Cs,Na)AlSi_2O_6 . nH_2O$]

(Not good agreement)

3670	650w
1630vw	591m, sh } b
1035vs, b	530–25? } b
(952)vs?	494m, sh } b
(817)s?	446–42s } b
767m } b	415–12sh } b
730m } b	
700s } b	

References 1, 9, 111, 548, 582

649
NEPHELITE
$KNa_3Al_4Si_4O_{16}$

ca. 2300vw	668sh
1650vw	582vw
ca. 1520w multiplet?	513m, sh
1410w	472s
1090–75sh	
1000–985vs, vb	
770sh	
736m, sh	
715–00m, b	
690m, b	

References 9, 33, 84, 111, 585

650
EUCRYPTITE
3[$LiAlSiO_4$]

1085b, sh	series of peaks on a vs, vb band	795w
1050m, sh	series of peaks on a vs, vb band	755w
1040inf	series of peaks on a vs, vb band	720inf
1015vs	series of peaks on a vs, vb band	705s, b
980vs	series of peaks on a vs, vb band	680m, sh
960vs	series of peaks on a vs, vb band	
940s	series of peaks on a vs, vb band	
915s, sh	series of peaks on a vs, vb band	

Reference 111

651
MILARITE
$[K_2Ca_4Be_4Al_2Si_{24}O_{60} . H_2O]$

3510
3420
1630–25vs, b?
1170inf }
1140–27vs }
1065inf } b
1030–12vs, b }
980–63sh }
798–95s } b
780s } doublet
750–40m, sh
695sh
599–90m
578–75m, sh
559–55s, sh
512–08m, sh
480–76sh }
462s } b triplet
435s }

References 9, 111, 172

652
DANBURITE
$4[CaB_2Si_2O_8]$

ca. 1140s } b doublet
1100s }
1030b, sh }
975s, sh }
950vs }
880s, sh } vb
821–10m, sh }
755–40sh }
698–80m
655–40s?
617?
540–25m
500–486s
455sh
444w
418m

References 9, 120, 533 (Xtal-reflection), *602, 614*

653
DUMORTIERITE
$4[(Al,Fe)_7BSi_3O_{18}]$

1650w, b
1460inf
1400–1370s, b, plat
1180sh
1050vs
1000–950b, sh
850b, sh
770s
720s
640b, sh
600vs, b
535s, sh
485s, b
435s, b

References 9, 119, 607

654
BAVENITE
$4[Ca_4(Be,Al)_4Si_9(O.OH)_{29} . (OH)_2]$

3520vw
3435vw
1131s, sh
1087s, sh
1037sh
1010vs
978sh
939sh
904sh
888s, sh
820sh
784sh
755s
711sh
653w, b
641b, sh
602w, b
565m
526m }
510sh } b
490m }
459s

Reference 172

655

TRIMERITE

24[Be(Ca,Mn)SiO_4]

985b, sh } b	560sh } vb
955sh } b	549s } vb
932vs } b	533sh } vb
810sh	515s, sh } vb
760s	490m, sh } vb
701s, sh } b	465s, sh } vb
688sh } b	445sh } vb
598vw	425sh } vb

Reference 10

656

PETALITE

4[$LiAlSi_4O_{10}$]

1210m, sh	558m, sh
1160sh	547s
1080vs	530s
1018m, sh	472s, sh
779m } b quartet	430s } b
757m } b quartet	415s } b
733m } b quartet	
706m } b quartet	

Reference 9

657

ORTHOCLASE

4[$KAlSi_3O_8$]

1150–38m, b, sh
1040vs, b
771–63m } b doublet
727–24m } b doublet
649–47s, sh
613sh
590–86s
542–40m, sh
515sh
477–60b, sh } b
439–30s } b

References 3, 9, 21, 35, 84, 111, 202, 301, 585

658

ADULARIA

var. Orthoclase

4[$KAlSi_3O_8$]

1150–30s, b, sh
1060–40vs, b
1000inf or sh
790–775sh
729–20m
649–40m
590–80s, b
540s, sh
440–32s

References 9, *17* and *18* (reflection), *35*, *84*, *111*, *202*, *301*, *397*

659
LENNILITE
nr$(MgAlFe)_6(Si,Al)_8O_{20}(OH)_4$
$. 9H_2O$

3680vw
3570m
3425m
1645w
1079w
997s
961w
815w
656m

Reference 168

660
HYALOPHANE
var. Orthoclase
$4[(K,Si,Ba,Al)AlSi_2O_8]$

1125b, sh } vb
1030vs, b }
730w, b
635sh
585m, b
545m, sh
430m, b
416sh

Reference 9

661
SANIDINE
$4[KAlSi_3O_8]$

1130–13m, b, sh } b
1090sh }
1060–26vs, b }
778–72b sh
728–20w, b
639–38m, sh
584s
546m, sh
463sh
430–23m, b

References 1, 9, 397

662
MICROCLINE
$4[KAlSi_3O_8]$

1640w
1325w
1140–28m, sh
1090–85sh
1055–45m, sh
1018–19vs
828
804–792sh
773–72w } b doublet
730–28w }
668sh?
650–45w
606sh
585m
545–33w, sp
467–65sh
430–27vs
380sh
330s, sh

References 1, 9, 33, 111, 201, 397

663
AMAZONITE
Var. Microcline
$4[KAlSi_3O_8]$

1135s } b
1110sh }
1075sh }
1055s, sh }
1010vs }
785sh
765w, b
725w, b

Reference 111

664
ANORTHOCLASE
$4[(Na,K)AlSi_3O_8]$

1115s, b, sh } b
1040vs, b }
775vw, vb
720vw, vb

Reference 111

665
ALBITE
$Ab_{100} . 4[NaAlSi_3O_8]$

1165–50m, b, sh	650–45s, sh
1133m, sh	610sh } b
1103–1095m, sh	587m } b
1045–28m, sh	533m
1010sh	470sh
1004–990vs	460m, sh
788–86m, sh } b quartet	425m
762–60w } b quartet	
745–42m, sh } b quartet	
722m, sh } b quartet	

References 1, 3, 4, 9, 17 (reflection), *21, 33, 84, 201, 397, 585, 602*

666
OLIGOCLASE
$Ab_{80}An_{20} . 4[(Na,Si,Ca,Al)AlSi_2O_8]$

1640w	643m, sh
1330vw	590m, b
1160m, b, sh } b doublet	538m, sh
1105–00m, b, sh } b doublet	470m, sh
1040vs } b triplet	430m
1010vs } b triplet	
995vs } b triplet	
790–88w } b quartet	
760w } b quartet	
748–40w } b quartet	
728–20w } b quartet	

References 1 (range of Ab : An ratios), *3, 9, 21, 33, 111, 397, 592*

667
ANDESINE
$Ab_{70-50}An_{30-50}$

1150–40sh } b
1110–1090b, sh } b
1030–20vs } b
1005vs } b
750–41s, b
633–32s, sh
585–80vs
543–37s, sh
465–62sh
430sh

References 1, 3, 9, 21, 111

668
LABRADORITE
$Ab_{50-30}An_{50-70}$

1140–32sh } b } vb
1101–00sh } b } vb
1028sh } vb
1000–995vs, b } vb
950b, sh } vb
778sh
750w
620sh } b
573m } b
535sh } b
468sh
422sh

References 1, 111

669
BYTOWNITE
$Ab_{30-10}An_{70-90}$. 8[(Na,Si,Ca,Al)-$AlSi_2O_8$]

1150–38m, sh ⎫
1102–1090m, sh ⎪
1027–10s ⎬ b
990sh ⎪
950b, sh ⎪
931–25vs ⎭
751–50sh
730s, sh
680sh
628–20s
582–73s
542–38s, sh
480–68m, sh
431–30sh
425–07s

References 1, 9, 111

670
ANORTHITE (An)
8[$CaAl_2Si_2O_8$]

1440w, b
1160–35s
1085–77m
1020–11b, sh ⎫ doublet ⎫
987b, sh ⎭ ⎪
950–46sh ⎬ vb
926vs, b ⎭
773sh
758–56w ⎫ b
728–27m, sh ⎭
668–62m, sh
624–20m ⎫
603sh ⎪ b
575–66m ⎬ triplet
540–36m ⎭
484–67w
470w, b
433sh

References 1, 3, 21, 33, 111, 201, 397, 602

671
MASKELYNITE
$(Na,Ca)(Al,Si)_4O_8$?
Variable composition

1087sh ⎫
1058sh ⎪
1032vs ⎬ vb
1000sh ⎪
962sh ⎪
932sh ⎭
787sh ⎫
750sh ⎪
718m ⎬ vb
668sh ⎪
638sh ⎭
612vw
582w

Reference 401

672
FAUJASITE
$Na_2CaAl_4Si_{10}O_{28}$. $20H_2O$

3580
3430
1660w, b
1150–20s, b
1100b, sh
1020vs, b
794m, sh ⎫
720m, vb ⎪
572m, vb ⎬ vb
505sh ⎪
455s, b ⎭

References 10, 111, 582

673

ANALCIME

$16[NaAlSi_2O_6 . H_2O]$

3620
1635–30w–m
1115–08sh } b
1040–20vs, b } b
972sh } b
862vw
775–66m } doublet
746–26m–s } doublet
666vw
635–20w } b
608sh } b
540sh
450–42s

References 1, 9, 111, 582

674

CHABAZITE

$2[CaAl_2Si_4O_{12} . 6H_2O]$

3440
1645m
1120–00b, sh } b
1030–25vs, b } b
910vw, b?
860vw, b?
700b, sh
517m } b triplet
465m } b triplet
415m } b triplet

References 9, 111, 582

675

GMELINITE

$(Na_2Ca)Al_2Si_4O_{12} . 6H_2O$

1655w
1080sh } vb
1040vs, b } vb
795sh } vb
775m } vb
740sh } vb
698m, sh } vb
600m, b } vb
522b, sh } vb
455s, b } vb
430sh } vb
415sh? } vb

References 10, 582

676

LEVYNE/LEVYNITE

$[NaCa_3Al_7Si_{11}O_{36} . 15H_2O]$

(approx.)
1110b, sh } b
1020vs } b
920vw } b
875vw, b

References 111, 582

677
THOMSONITE
$4[NaCa_2Al_5Si_5O_{20} . 6H_2O]$

3550
3440
1640w, b
1150–00b, sh ⎫
1050sh ⎪
1008–00vs ⎬ vb
940sh ⎪
920sh ⎭
780sh
730sh
680vb, sh
595m, vb
440s, b

References 9, 111, 582

678
GONNARDITE
$2[Na_2CaAl_4Si_6O_{20} . 7H_2O]$

(approx.)
1235sh ⎫
1100b, sh ⎬ b
1040vs, b ⎭

References 111, 582

679
ERIONITE
$NaKCaAl_4Si_{12}O_{32} . 12H_2O$

(approx.)
1160sh ⎫
1060vs, b ⎭ b
990vw, b

References 111, 582

680
FERRIERITE
$(NaK)_4Mg_2(Al_6Si_{30}O_{73} . 20H_2O)$

(approx.)
1220sh ⎫
1150w, sh ⎬ b
1060vs, b ⎭
870w

References 111, 582

681
EDINGTONITE
$2[BaAl_2Si_3O_{10} . 4H_2O]$

1165w, sh ⎫
1100sh ⎪
1050sh ⎬ b
1020vs ⎪
990sh ⎪
975sh ⎭
875vw, b

Reference 111

682
GISMONDINE(ITE)
$8[CaAl_2Si_2O_8 . 4H_2O]$

3550
3440
3250
1655w, b
1130–10b, sh ⎫
1027–10vs, b ⎪
785sh ⎪
735sh ⎪
685vb, sh ⎬ vb
597m, vb ⎪
585sh ⎪
465sh ⎪
440s, b ⎪
415sh? ⎭

References 10, 111

683
NATROLITE
$8[Na_2Al_2Si_3O_{10} . 2H_2O]$

3550–00s, b
3330m
3240b, sh
1640vw, b
1090sh
1067m, sh
1020–980vs, vb
680b, sh
600b, sh
545m, sh
515m, b, sh
417s

References 9, 36, 111, 582

684
SCOLECITE
$8[CaAl_2Si_3O_{10} . 3H_2O]$

3585
3503
3405
3390
3320
3230
1667m
1656sh?
1595vw
1102s, sh ⎫
1068sh ⎪
1045sh ⎪
1025m, sh ⎬ vb
991vs ⎪
954m, sh ⎪
932m, sh ⎭
718m, sh
697m, sh
677sh
635s
605b, sh
520sh
500m, b
450sh
428s
415sh

References 10, 111, 582

685

MESOLITE

$8[Na_2Ca_2Al_6Si_9O_{30} . 8H_2O]$

(approx.)
1090sh }
1065s }
1040sh }
1005inf } b
990vs }
965sh }
870vw, b

References 111, 582

686

LAUMONTITE

$4[CaAl_2Si_4O_{12} . 4H_2O]$

3560	765s
3470	625sh
1655m	565s, sh
1134s, sh } b	525m, sh
1095sh } b	492m, sh
1038vs } b	432s, b
1000m, sh } b	410sh
960m, sh } b	

References 10, 111, 582

687

PHILLIPSITE

$(K_2Na_2Ca)Al_2Si_4O_{12} . 4\frac{1}{2}H_2O$

3440
1640w
1160–1080vb, sh } b
1050–25vs, b } b
780sh
680vb, sh
630sh?
590m, b
445s
417sh

References 9, 110, 582

688

WELLSITE

$nr(Ba,Ca,K_2)Al_2Si_3O_{10} . 3H_2O$

(approx.)
1130b, sh } vb
1030vs } vb

References 111, 582

689
HARMOTOME
$2[BaAl_2Si_6O_{16} . 6H_2O]$

(approx.)
1140sh }
1110sh } b
1030vs }
870vw, vb

References 111, 582

690
STILBITE
$2[NaCa_2Al_5Si_{13}O_{36} . 14H_2O]$

3440
3250
1645w
1145m, b, sh }
1050–30vs, b } b
870vw }
780sh
710vb, sh
600sh
560m, vb
440m, b
415sh?

References 9, 111, 582

691
STELLERITE
$[Ca_2Al_2Si_{14}O_{36} . 14H_2O]$

(approx.)
1160b, sh }
1050vs } b
990vw, b

References 3, 111, 582

692
EPISTILBITE
$CaAl_2Si_6O_{16} . 5H_2O$

(approx.)
1185vb, sh }
1060vs } b
880vw, b

References 111, 582

693
BREWSTERITE
$(Sr,Ba,Ca)Al_2Si_6O_{16} . 5H_2O]$

3605
3550
1640
1200sh } vb
1150–40b, sh } vb
1060sh } vb
1135–15vs, b } vb
880vw, b
770sh } vb
730sh } vb
703m, vb } vb
635sh } vb
595m, vb } vb
538m, sh } vb
460–45s, b } vb
407m, sh } vb

References 10, 111, 582

694
HEULANDITE
$[(Na,Ca)_{4-6}Al_6(Al,Si)_4Si_{26}O_{72} . 24H_2O]$

3500–3450w, b
1635–15w
1205–1195sh } b
1060–40vs, b } b
ca. 885w, b
800w
780sh
720m, sh
670m, sh
602m, b
520sh
465s, b
415sh

References 9, 36, 111, 347, 582

695
DACHIARDITE
$[(Ca,Na_2K_2)_3Al_4Si_{18}O_{45} . 14H_2O]$

(approx.)
1205b, sh } b
1050vs, b } b
890vw, vb

Reference 111

696
LAUBANITE
$Ca_2Al_2Si_5O_5 . 6H_2O$

(approx.)
1050vw, sh } vb
1030vs, vb } vb
990inf } vb
880vw, vb

Reference 111

697
SODALITE
$2[Na_4Al_3Si_3O_{12}(Cl)]$

1030–20sh
985–80vs
740–37s, sp } doublet
715–12m, sp, sh } doublet
670m–s
475sh
465s } b doublet
440–38s } b doublet

References 3, 9, 21, 35, 111, 590, 582

698
NOSEAN/NOSELITE
$[Na_8Al_6Si_6O_{24} . (SO_4)]$

1630 } b
1145–10b, doublet sh } b
1010–05vs } b
728–20b, sh
702–695w
660w
617b, sh
465sh
455s
430sh

References 9, 111

699
HAUYNE/HAUYANITE
$[(NaCa)_{4-8}Al_6Si_6O_{24}(SO_4)_{1-2}]$

1200sh } b
1150–00vb sh } b
1010vs, b } b
728–20b, sh
702–695m
656m, b
616m, sh
540sh
452s, b
430sh?

References 9, 111

700
LAZURITE
$(Na,Ca)_8(Al,Si)_{12}O_{24}(SO_4,\text{'S'}_n)]$

1110–1088s, vb, sh } vb
1056sh } vb
1016–10vs, b } vb
970m, sh } vb
930m, sh } vb
904sh } vb
880w? } vb
730–28sh
695–85m, b, sh
647m
620sh
570m
510sh } b
470sh } b
445s } b

References 9, 111

701
HELVINE
$[(Mn,Fe,Zn)_8Be_6Si_6O_{24}S_2]$

1099
1065sh
1000–980sh
955–49vs } b, sh } b
930–20vs
914sh ?
826
775–70s, sh
745–42m, sh } b
711–05s
535vs, b
424vw

References 9, 111, 172

702
DANALITE
$[(Fe,Zn,Mn)_8Be_6Si_6O_{24}S_2]$

1099
980sh
950–47vs
925–23vs
910–08sh
775–71m
745w
715–11m
532
469

References 111, 172

703
MARIALITE (Ma)
$2[Na_4Al_3Si_9O_{24}Cl]$

1190m
1110b, sh } b
1030vs, b
995inf
760w

Reference 111

704
WERNERITE
$(Na,Ca,K)_4Al_3(AlSi)_3Si_6O_{24}$-
(Cl, CO_3,SO_4, OH)?

1540–30w
1430w
1180–75s, sh
1108–00b, sh } b
1020–00vs, b
880sh
755–40m, sh } b
700w
613–10m–s
545–40m–vs
462–50s, sh
415m, b, sh

References 9, 111, 360

705
MEIONITE
$2[Ca_4Al_6Si_6O_{24}(SO_4,CO_3,Cl_2)]$

(approx.)
1160sh } b
1070b, sh } b
995vs } b

Reference 111

706
CANCRINITE
$2[Na_3CaAl_3Si_3O_{12}CO_3]$?

1638w
1570vw
1505w
1480sh
1438vw
1400–1390w
1120–15m, sh } b
1055–38s, sh } b
1008–00vs, b } b
945sh } b
860–55vw } b
748m, sh } b
690–685m, b
625m, b
570m
530sh
500sh
465s } b doublet
430s } b doublet

References 9, 35, 111

707
DAVYNE
$[(Na,K,Ca)_{6-8}Al_6Si_6O_{24}$-
$(SO_4CO_3Cl_2)_{1-2}]$

1510w
1412–1390vw
1155–32s, sh } vb
1055–50sh } vb
1025–15vs, b } vb
960sh } vb
860sh } vb
683m, b
613m
565m
440s, b
430sh?

References 10, 111

708
CYMRITE
$64[BaAlSi_3O_8OH]$

(approx.)
3450w
1615vw
1080b, sh
1020s } b doublet
1000s } b doublet
934sh
901m

Reference 234

709
DATOLITE
$4[CaBOH \cdot SiO_4]$

3490 (vb)	578m (b)
1160s, sh (vb)	550sh (b)
1102s, sh (vb)	533m, sh (b)
1050vs (vb)	505s (b)
1015s, sh (vb)	478m, sh (b)
965sh (vb)	455sh (b)
955vs (b; vb)	425m (b)
930vs (b; vb)	
885s, sh (vb)	
853m, sh (vb)	
788s, sp (vb)	
692m (vb)	

References 9, 120, 631

710
GADOLINITE
$2[Y_2FeBe_2Si_2O_{10}]$

1180sh (vb)
1100inf (vb)
1035s, b (vb)
970sh (vb)
802vw
782sh
700vw
480sh (vb)
465sh (vb)
450s (vb)

References 10, 226

711
HOMILITE
$(Ca,Fe)_3B_2Si_2O_{10}$

3420
1635w, b
1410m, vb, sh
1000vs, vb
750m, b, sh
535b, sh
475s, vb

Reference 10

712
LABRADORITE/ANEMOUSITE
$Ab_{30-50}\ An_{70-50}$

(approx.)
1175inf (b)
1090s, sh (b)
1010vs, b (b)

Reference 111, see also p. 224

Also	*Reference*
Wairakite $8[CaAl_2Si_4O_{12} \cdot 2H_2O]$	*582*
Haydenite $2[CaAl_2Si_4O_{12} \cdot 6H_2O]$	*582*
Mordenite $nr[(Ca,Na_2K_2)_4Al_8Si_{40}O_{96} \cdot 28H_2O]$	*582*
Pumpellyite $2[Ca_4(Al, Fe^{III}, Fe^{II}, Mg)_6Si_6O_{23}(OH)_32H_2O]$	*3*

UNCLASSIFIED SILICATES

713
BUDDINGTONITE
(a feldspar)
$NH_4AlSi_3O_8 \cdot \frac{1}{2}H_2O$

3419m?
1117m, sh } b
1034vs } b
785w
731sh
711m
621sh
583m
546sh
466sh
425m

Reference 252

714
JENNITE
$Na_2Ca_8(SiO_3)_3Si_2O_7$

3565s
3490m, sh
3170m, vb
1630w
ca. 1460w
1070s, sh } b
ca. 975vs } b
905m, sh
740w
710s

Reference 378

715
MAGADIITE
$NaSi_7O_{13}(OH)_3 . 3H_2O$

(approx.)
1640w
1220sh
1120sh
1080vs
ca. 750w, b, doublet
700vw
ca. 600w, doublet

Reference 461

716
SKOLEZITE
$8[CaAl_2Si_3O_{10} . 3H_2O]$

3585m, sh
3503m
3405m, b } b
3390sh
3320m, sh
3230m
1667w
1656sh
1595vw
1102s, sh
1068sh
1045sh } b
1025s, b, sh } vb
991vs, sp
954s, sh
932s, sh
718s, sh
697s, sh
677b, sh
635m, b
605b, sh } vb
520sh
500m, b
450sh
428s
415sh

Reference 10

717
SORENSENITE
$Na_4SnBe_2Si_6O_{16}(OH)_4$

ca. 3550w
1700vw
1100m, sh
1050sp, sh
960s
880m, sh
800m, sp
790sh
700w
640sh
600m, sh
580w, sh
530sh
470sh
440m, sh
410vs

Reference 388

	References		References
Azalmatolite	*18* (reflection)	Reedmergnerite $NaBSi_3O_8$	*614*
Andalanite	*17* (reflection)	Leucosphenite $Na_3CaBaTi_3Si_9O_{29}$	*614*
Agalmatolite	*18* (reflection)		
Bayalite	*18* (reflection)		
Miloschite	*480*		
Deweylite	*127*		

METASILICATES

718
POTASSIUM METASILICATE (*S*)
K_2SiO_3

- 3330m
- 1625vw
- ca. 1100b, sh
- 990vs, vb
- ca. 840b, sh
- 770vw, b
- 730m, sh
- 620w, b
- 470s } vb
- 455sh } vb

References 7, 32

719
SODIUM METASILICATE (*S*)
$Na_2SiO_3 . 5H_2O$

3400s } doublet	775m
3280s } doublet	720–15m, b
2330w, b	667w
1695w	583m
1165w	545s
1125–17w, b	534sh
1043s?	463–52m–s, b
1007m, sh?	
980–76vs	
898s, sh	
873vs?	
832m	

References 7, 32, 79

720
LITHIUM METASILICATE (*S*)
Li_2SiO_3

1070–55b, sh } vb	450–47m, b
980–75vs } vb	410m
950–42s, sh } vb	390sh
879sh } vb	380sh
855–47vs } vb	350sh
736–35m	
605m	
587–80vw, sh	
545m, sh } b	
521–20m } b	
500m, sh } b	

References 4, 79

721
LITHIUM SILICATE (*S*)
$Li_2Si_2O_5$

1220s, sh } vb	555m, sh } vb
1110m, sh } vb	535sh } vb
1025vs, b } vb	475s } vb
960sh } vb	385sh } vb
945vs } vb	330m, sh } vb
790m } b doublet	265m, sh } vb
760m } b doublet	
730sh	
640w	

Reference 4

URANIUM MINERALS

722
BOLTWOODITE
$K_2(UO_2)_2(SiO_3)_2(OH)_2 \cdot 5H_2O$

3400m } b
3125m }
1640w
1540w
1430vw
1335vw
1085sh }
990s }
935s } b
875sh }
860vs, b }
790b, sh }

Reference 87

723
COFFINITE
$USiO_4$

(approx.)
ca. 3390w, b
1620vw
1180m, b, sh } vb
ca. 900vs, b }

References 89, 103

724
GASTUNITE
$(K,Na)_2(UO_2)_3(Si_2O_5)_4 \cdot 8H_2O$

(approx.)
ca. 3550m, b
1610w
1110s, b, sh }
1020vs, b } b
920vs }
870s, sh }
770w, b

Reference 102 (inc. NH_4 and K vars.)

725
HAIWEEITE
$CaU^{IV}{}_2Si_6O_{17} \cdot 5H_2O$

(approx.)
3450w
1640w
1180sh }
1080s, vb } vb
1000b, sh }
910s, b }
780m, b

Reference 54

726
RANQUILITE
$Ca_3(UO_2)_4Si_{10}O_{27} . 24H_2O$

(approx.)
3450m
ca. 1640w
1130sh
1100s, b ⎫
990s, b ⎬ vb triplet
910s, b ⎭
780w, b

Reference 54

727
URANOPHANE/α-URANOTILE
$Ca(UO_2)_2Si_2O_7 . 6H_2O$

3500	778m, sh
3440	670w
3150	555m, b ⎫
ca. 1640m, b	485m, b ⎬ vb
ca. 1150m, b, sh	450m, sh ⎭
1102sh	
1004s ⎫	
940s ⎬ b triplet	
860vs ⎭	

References 9, 54, 116

ORGANIC MINERALS

728
FICHTELITE
$C_{19}H_{34}$

3003sh
2990sh
2980sh
2966m, sh
2950sh
2940sh
2925vs
2895sh
2870s, sh
2850
2845
2840
(3003–2840: vb; 2850, 2845, 2840: triplet sh)
1470sh
1450m
1385m
1368sh
1348w
1338w
1323w
1308w
1268w
1258w
1228w
1200w
1186w
1145w
1115w
1017w
980w
970w
952w
938w
920w
890w
872w
855w
836w
818w

Reference 113

729
BERNSTEIN/AMBER
$(CH_2COOH)_2$

2990sh
2980sh
2966sh
2940sh
2925s
2870m, sh
2850sh
1730s
1705s
(1730, 1705: b)
1640sh
1450m
1375m
1260sh
1158m
(1260, 1158: b)
1020m, sh
1000sh
890vw

References 10, 113

730
HUMBOLDTINE (*S*)
$Fe(C_2O_4) . 2H_2O$

3340s, vb
1625vs, b
1360w, sp
1316m, sp
822m
730m, b
590b, sh } b
495s } b

References 10, 113

731
MELLITE
$16[Al_2\{C_6(COO)_6\} . 18H_2O]$

3150?
1700s, sh } b
1560vs, b } b
1445vs, sh } b
1390sh } b
1348vs, sh } b
970m, sh } b
920sh } b
902m } b
800m, b } b
760sh } b
740m, sh } b
655sh } b
642m } b
610m, b } b
507w

References 10, 113

732
OXAMMIT (*S*)
$(NH_4)_2(C_2O_4) . H_2O$

3200 } b
3100 } b
3010 } b
2870 } b
1740sh } b
1702sh } b
1640b, sh } b
1600s } b
1462sh } b
1440s, b } b
1430sh } b
1410sh } b
1310s, b
805m } doublet
775m } doublet
722m
642m, b
525vw
430m, b

References 10, 113

733
WHEWELLITE
$Ca(C_2O_4) . H_2O$

3490w
3430–00w
3350–00w
3270sh
3080w
1660sh
1630vs
1470sh
1385–80sh
1325–20s, sp
955–45w
892–80w
788–85s, sp
670–65m, b
605b, sh
520m, b
427vw

References 10, 33, 113

734
ASPHALT

(approx.)
3450m
2900s
1610w, b
1460m
1380w

Reference 89

735
ASPHALTITE (Group name)/ BITUMEN

(approx.)
3450s
ca. 2900w
1640w
1450vw
1390vw
1100w, b
1020m, vb

References 89, 620

736
ALBERTITE

(approx.)
3390m
2900vs
1695m
1585m, sh
1450s
1370m
970w
725m

Reference 511

737
'OIL'

1610vw
1470s
1380m
1310vw
1165vw
810vw
765w, sh
740w, sh
720w
610vw
585vw

Reference 201

738
EXINITE

ca. 3700sh
ca. 3500b, sh
3200m, sh
3000vs, sp
2800sh
1690sh
1600vs, b
1450vs
ca. 1300inf
ca. 1200s, vb
1100sh
1040sh
980sh
880m } part resolved doublet
820m }
740m
710sh

Reference 136

739
'VITRINITE'

ca. 3500m, b
3150m, sh
3000s
2900sh
1600vs, b
1460s
ca. 1270s, b
1220sh
1150inf
1050sh
880m } part resolved triplet
820m }
740m }

Reference 136

PART IV

References

References

1. Lyon, R. J. P., 'Evaluation of Infrared Spectrophotometry for Compositional Analysis of Lunar and Planetary Soils', Stanford Research Institute. Final report under contract NASr-49(04). Published by N.A.S.A. as Technical Note D—1871 (1963)
2. Lyon, R. J. P., 'Infrared Absorption Spectroscopy', in *Physical Methods in Determinative Mineralogy*, Chapter 8 (ed. J. Zussman), Academic Press, New York and London (1967)
3. Omori, K., Science Reports, 3rd Ser., Vol. 9, No. 1, p. 65, Tohoku University (1964)
4. Nyquist, R. A. and Kagel, R. O., *Infrared Spectra of Inorganic Compounds*, Academic Press (1971)
5. Achmanova, M. V., *Usp. Khim. S.S.S.R.*, **28** (3), 312 (French translation distributed by U.S. Dept. of Commerce, Offices of Technical Services) (1959)
6. *Inorganics and Related Compounds*, Philadephia, Sadtler Research Laboratories Inc.
7. Miller, F. A. and Wilkins, C. H., *Analyt. Chem.*, **24,** 1253 (1952)
8. Nakamoto, K., *Infrared Spectroscopy of Inorganic and Co-ordination Compounds*, 2nd ed., New York, Wiley Interscience (1970)
9. Moenke, H., *Mineralspecktren*, Academie Verlag, Berlin (1962)
10. Moenke, H., *Mineralspecktren II*, Academie Verlag, Berlin (1966)
11. Karr, Jr. C. and Kovach, J. J., *Appl. Spectrosc.*, **23** (3), 219 (1969)
12. Ferraro, J. R., *Low Frequency Vibrations of Inorganic and Co-ordination Compounds*, Plenum Press, New York (1971)
13. *I.U.P.A.C. Tables of Wavenumbers for the Calibration of Infrared Spectrometers*, Butterworths, London (1961)
14. Stanfield, J. E., Shepperd, D. E. and Harrison, H. S., *Laboratory Methods in Infrared Spectroscopy*, Chap. 2 (ed. Miller, R. G. J.), London, Heyden and Sons (1965)
15. Smallwood, S. E. F. and Hart, P. B., *Spectrochim. Acta*, **19,** 285 (1963)
16. Ulrich, W. F., 'Applications of Specular Infrared Reflectance', in *The Analyzer*, p. 14 (Beckman Inc. house journal) (1961)
17. Schaeffer, Cl., Matossi, F. and Wirtz, K., *Z. Phys.*, **89** (3 and 4), 210 (1934)
18. Matossi, F. and Kruger, H., *Z. Phys.*, **99** (1 and 2), 1 (1936)
19. Serratosa, J. M. and Bradley, W. F., *J. phys. Chem.*, **62** (10), 1164 (1958)
20. Serratosa, J. M., Hidalgo, A. and Vinas, J. M., *Nature*, **195,** 486 (1962)
21. Omori, K., Science Reports, Vol. 7, p. 101, Tohoku University (1961)
22. Gadsden, J. A., Smith, W. L. and Parker, J., *Atmosph. Environ.*, **4,** 667 (1970)
23. Lyon, R. J. P., Tuddenham, W. M. and Thompson, C. S., *Econ. Geol.*, **54,** 1047 (1959)
24. Duncan, A. B. F., *Chemical Applications of Spectroscopy*, Interscience, New York (1956)
25. Wilson, E. B., Decius, J. C. and Cross, P. C., *Molecular Vibrations—the Theory of Infrared and Raman Spectra*, McGraw-Hill, London (1955)
26. Hertzberg, G., *Infrared and Raman Spectra of Polyatomic Molecules*, Van Nostrand, New York (1945)
27. Muller, L. D., 'Laboratory Methods of Mineral Separation', in *Physical Methods in Determinative Mineralogy*, Chap. 1 (ed. J. Zussman), Academic Press, London and New York (1967)
28. Lyon, R. J. P. and Burns, E. A., *Econ. Geol.*, **58,** 274 (1963)
29. Szymanski, H. A. and Peller, P., *Appl. Spectrosc.*, **14** (7), 107 (1960)
30. Raman, C. V., *Proc. Indian Acad. Sci.*, **26A** (6), 339
31. Louisfert, J., *Comptes Rendus hebd. Séanc Acad. Sci. Paris*, **248,** 1150 (1959)
32. Miller, F. A., Carson, G. L., Bentley, F. F. and Jones, W. H., *Spectrochim. Acta*, **16,** 135 (1960)
33. Hunt, J. M., Wisherd, M. P. and Bonham, L. C., *Analyt. Chem.*, **22,** 1478 (1950)
34. Hidalgo, A. and Serratosa, J. M., *Ion*, **16,** 645 (1956)
35. Launer, P. J., *Am. miner.*, **37,** 764 (1952)
36. Keller, W. D., Spotts, J. H. and Biggs, D. L., *Am. J. Sci.*, **250,** 453 (1962)
37. Duval, R., Duval, C. and Lecomte, J., *Bull. Soc. chim. fr. Mem.*, 5ᵉ serie, **10** (11 and 12), 374 (1943)
38. Setkina, O. N., *Zap. vses miner. Obshch.*, **88,** 39 (1959)
39. Hunt, J. M. and Turner, D. S., *Analyt. Chem.*, **25,** 1169 (1953)
40. Garcia, S. G., Beutelspacher, H. and Flaig, W., *Annls. Real Soc. espan. fis. Quim.*, **528,** 369 (1956)
41. Lyon, R. J. P. and Tuddenham, W. M., *Nature*, **185,** 835 (1960)
42. Lyon, R. J. P. and Tuddenham, W. M., *Nature*, **185,** 374 (1960)
43. Tarte, P., *Nature*, **186,** 234 (1960)
44. Tarte, P. and Ringwood, A. E., *Nature*, **193,** 971 (1962)
45. Fripiat, J. J. and Toursaint, F., *Nature*, **186,** 627 (1960)
46. Hexter, R. M., *J. opt. Soc. Am.*, **48,** 770 (1958)
47. Corbridge, D. E. C. and Lowe, E. J., *J. chem. Soc.*, Pt. 1, 493 (1954)
48. Corbridge, D. E. C., *J. appl. Chem.*, **6,** 456 (1956)
49. Roy, D. M. and Roy, R., *Geochim. cosmochim. Acta*, **11,** 72 (1957)

50. Plyusnina, E. E., *Zh. strukt. Khim.*, **2** (3), 330 (1961)
51. Lazarev, A. N. and Tenisheva, T. F., *Optika Spektrosk.*, **12** (2), 215 (1962)
52. Lazarev, A. N., *Kristallografiya*, **6** (1), 125 (1961)
53. Hidalgo, A. and Serratosa, J. M., *An. Real Soc. espan. fis. Quim.*—Serie B Quimica, **52B**, 101 (1956)
54. De Mandirola, O. B. and Silberman, E., *An. Asoc. quim. argent.*, **49**, 293 (1961)
55. Waldron, R. D., *Phys. Rev.*, **99** (6), 1727 (1955)
56. Narayanan, P. S. and Lakshmanan, B. R., *J. Indian Inst. Sci.*, Sections A and B, **40** (1), 1 (1958)
57. Louisfert, J. and Pobeguin, T., *Comptes Rendus hebd. Séanc. Acad. Sci. Paris*, **235**, 287 (1952)
58. Hidalgo, A., Serratosa, J. M. and Jubrias, M., *An. Edafol. Fisiol. vegl.*, **15**, 607 (1956)
59. Weir, C. E. and Lippincott, E. R., *J. Res. natn. Bur. Stand.*—A. Physics and Chemistry, **65A** (3), 173 (1961)
60. Brame, Jr., E. G., Margrave, J. L. and Meloche, V. W., *J. inorg. nucl. Chem.*, **5** (1), 48 (1957)
61. Caillere, S. and Pobeguin, T., *Bull. Soc. fr. miner. Cristallogr.*, **33** (1–3), 36 (1960)
62. Van der Marel, H. W. and Zwiers, J. H. L., *Silic. Ind.*, **24** (7–8), 359 (1959)
63. Caillere, S. and Pobeguin, T., *Bull. Soc. fr. Miner. Cristallogr.*, **35** (1), 48 (1962)
64. Titeica, R. and Palade, G., *Revue Phys.* (Academie de la Republique populaire Roumaine), **4** (1), 93 (1959)
65. Adler, H. H. and Kerr, P. F., *Am. Miner.*, **48**, 124 (1963)
66. Dayal, B., *Proc. Indian Acad. Sci.*, **32** (5), 304 (1950)
67. Duval, C. and Lecomte, J., *Bull. Soc. chim. Fr.*, Jan.–Feb., 101 (1952)
68. Meloche, V. W. and Kalbus, G. E., *J. inorg. nucl. Chem.*, **6** (2), 104 (1958)
69. Rocchiccioli, C., 'Groupement pour l'Avancement des Methodes d'Analyse Spectrographie des Produits Metallurgigue', 20th Congress, p. 105 (1957)
70. Cabannes-Ott, C., *Comptes Rendus hebd. Séanc. Acad. Sci. Paris*, **244** (20), 2491 (1957)
71. Mara, R. T., Sutherland, G. B. B. M. and Tyrell, H. V., *Phys Rev.*, **96**, 801 (1954)
72. Vratny, F., *Appl. Spectrosc.*, **13** (3), 59 (1959)
73. Moenke, H., *Naturwissenschaften*, **49** (1), 7 (1962)
74. Tai, H. and Underwood, A. L. *Analyt. Chem.*, **29**, 1430 (1957)
75. Tarte, P., *Spectrochim. Acta*, **18**, 467 (1962)
76. Pustinger, Jr., J. V., Cave, W. T. and Nielsen, M. L., *Spectrochim. Acta*, **15**, 909 (1959)
77. Addison, C. C. and Gatehouse, B. M., *Chemy Ind.*, 464 (1958)
78. Lazarev, A. N. and Tenisheva, T. F., *Optics Spectrosc.*, **11**, 316 (1961)
79. Lazarev, A. N. and Tenisheva, T. F., *Optics Spectrosc.*, **10**, 37 (1961)
80. Lazarev, A. N., *Optics Spectrosc.*, **9**, 103 (1960)
81. Kolesova, V. A. and Ryskin, Ia. I., *Optics Spectrosc.*, **7**, 165 (1959)
82. Godnev, I. N., Aleksandrovskaia, A. M. and Rigina, I. V., *Optics Spectrosc.*, **7**, 172 (1959)
83. Ignat'eva, L. A., *Optics Spectrosc.*, **6**, 527 (1959)
84. Kolesova, V. A., *Optics Spectrosc.*, **6**, 38 (1959)
85. Pastiniak, I., *Optics Spectrosc.*, **6**, 64 (1959)
86. Glemser, O., *Nature*, **183**, 943 and 1476 (1959)
87. Honea, R. M., *Am. Miner.*, **46**, 12 (1961)
88. Stubican, V. and Roy, R., *Am. Miner.*, **46**, 32 (1961)
89. Abdel-Gawad, A. M. and Kerr, P. F., *Am. Miner.*, **46**, 402 (1961)
90. Meyer, J. W. and Jaunarajs, K. L., *Am. Miner.*, **46**, 913 (1961)
91. Grim, R. E. and Kulbicki, G., *Am. Miner.*, **46**, 1329 (1961)
92. Whelan, J. A. and Goldich, S. S., *Am. Miner.*, **46**, 1412 (1961)
93. Steele, W. C. and Decius, J. C., *J. chem. Phys.*, **25**, 1184 (1956)
94. Jones, L. H., *J. chem. Phys.*, **25**, 1069 (1956)
95. Stubican, V. and Roy, R., *J. Am. Ceram. Soc., Ceramic Abstracts*, **44**, 625 (1961)
96. Barraclough, C. G., Lewis, J. and Nyholm, R. S., *J. chem. Soc.*, Pt. IV, 3552 (1959)
97. Addison, C. C. and Gatehouse, B. M., *J. chem. Soc.*, 613 (1960)
98. Anbar, M., Halmann, M. and Pinchas, S., *J. chem. Soc.*, 1242 (1960)
99. Brindley, G. W. and Zussman, J., *Am. Miner.*, **44**, 185 (1959)
100. Faust, G. T., Hathaway, J. C. and Millot, G., *Am. Miner.*, **44**, 342 (1959)
101. Muto, T., Meyrowitz, R., Pommer, A. M. and Musano, T., *Am. Miner.*, **44**, 633 (1959)
102. Honea, R. M., *Am. Miner.*, **44**, 1047 (1959)
103. Fuchs, L. H. and Hoekstra, H. R., *Am. Miner.*, **44**, 1057 (1959)
104. Anderson, W. R., Stringham B. and Whelan, J. A., *Am. Miner.*, **47**, 1303 (1962)
105. Sun, M. S., *Am. Miner.*, **47**, 1453 (1962)
106. Hausen, D. M., *Am. Miner.*, **47**, 637 (1962)
107. Adler, H. H. and Kerr, P. F., *Am. Miner.*, **47**, 700 (1962)

108. Stringham, B., *Am. Miner.*, **47,** 1187 (1962)
109. Omori, K. and Konno, H., *Am. Miner.*, **47,** 1191 (1962)
110. Huang, C. K. and Kerr, P. F., *Am. Miner.*, **45,** 311 (1960)
111. Milkey, R. G., *Am. Miner.*, **45,** 990 (1960)
112. Serratosa, J. M., *Am. Miner.*, **45,** 1101 (1960)
113. Moenke, H., *Chemie Erde*, Band 21, left 2, 239 (1961)
114. Allpress, J. G. and Hambly, A. N., *Aust. J. Chem.*, **12,** 569 (1959)
115. Moenke, H., *Janaer Jahrbuch*, Pt I, 239 (1961)
116. Moenke, H., *Janaer Jahrbuch*, Pt II, 402 (1960)
117. Moenke, H. and Moenke-Blankenburg, L. *Janaer Jahrbuch*, Pt II, 396 (1960)
118. Moenke, H., *Janaer Jahrbuch*, Pt II, 361 (1959)
119. Moenke, H., *Silikat Technik*, 323 (July 1961)
120. Moenke, H., *Silikat Technik*, 287 (Aug. 1962)
121. Tarte, P., *Spectrochim. Acta*, **19,** 25 (1963)
122. Tarte, P., *Spectrochim. Acta*, **19,** 49 (1963)
123. Brown, D. H., *Spectrochim. Acta*, **19,** 585 (1963)
124. Newman, G. and Powell, D. B., *Spectrochim. Acta*, **19,** 213 (1963)
125. Benoit, A., *Spectrochim. Acta*, **19,** 386 (1963)
126. Baron, G., Caillere, S., Lagrange, R. and Pobeguin, T., *Comptes Rendus hebd. Séanc. Acad. Sci. Paris*, **245,** 92 (1957)
127. Adler, H. H., Bray, E. E. *et al.*, Preliminary Report No. 8, API Project 49. Columbia University, New York (July 1950)
128. Tarte, P., *Nature*, **191,** 1002 (1961)
129. Szymanski, H. A. and Porinelli, R., *Nature*, **191,** 64 (1961)
130. Matossi, F. and Bronder, O., *Z. Phys.*, **III,** Pts 1 and 2 pl. (1938)
131. Duval, C. and Lecomte, J., *Comptes Rendus hebd. Séanc. Acad. Sci., Paris*, **240,** 66 (1955); also *Mickrochimica Acta*, 454 (1956)
132. Dasent, W. E. and Waddington, T. C., *J. chem. Soc.*, 1429 (1960)
133. Keeling, P. S., *Trans. Brit. Ceramic Soc.*, **62,** 549 (1963)
134. Hafner, S. and Laves, F., *Z. Kristallogr.*, **115,** 321 (1961)
135. Rocchiccioli, C., *Comptes Rendus hebd. Séanc. Acad. Sci., Paris*, **244,** 2704 (1957)
136. Bent, R. and Brown, J. K., *Information Circular No. 243*, British Coal Utilization Research Association
137. Paillette, P., Lankoande, A., Hadni, A. *et al. Bull. Soc. chim. Fr.*, **88,** 432 (Mar. 1960)
138. Tsuboi, M., Terada, M. and Shimanouchi, T., *J. chem. Phys.*, **36** (5), 1301 (1964)
139. Braithwate, R. S. W. and Ryback, G., *Mineralog. Mag.*, **33,** 441 (1963)
140. Scholze, H. and Dietzel, A., *Naturwissenschaften*, **42,** 575 (Nov. 1955)
141. Underwood, A. L., Toribara, T. Y. and Neuman, W. F., *J. Am. Chem. Soc.*, **77,** 317 (1955)
142. Dachille, F. and Roy, R., *J. Am. Ceram. Soc.*, **42,** 78 (1959)
143. Pobeguin, T., *Acad. Sci.*, 3289 (28 Mar. 1960)
144. Pobeguin, T. and Lecomte, J., *Comptes Rendus hebd. Séanc. Acad. Sci., Paris*, **236,** 1544 (1953)
145. Takeuchi, Y., *Mineralog. J.*, **2,** 245 (1958)
146. Guerchais, J. E., Haumesser, A. and Rhomer, R., *Comptes Rendus hebd. Séanc. Acad. Sci., Paris*, **260,** 5571 (1965)
147. Buchanan, R. A., *J. chem. Phys.*, **31,** 870 (1959)
148. Duval, C. and Lecomte, J., *Z. Electrochemie*, **60,** 582 (1956)
149. Duval, C. and Lecomte, J., *Comptes Rendus hebd. Séanc. Acad. Sci., Paris*, **227,** 1153 (1948)
150. Duval, C. and Lecomte, J., *Bull. Soc. chim. Fr.*, 101 (1947)
151. Duval, C. and Lecomte, J., *Bull. Soc. chim. Fr.*, 106 (1947)
152. Duval, C. and Lecomte, J., *Bull. Soc. chim. Fr.*, 276 (1947)
153. Beutelspacher, H., *LandbForsch.-Völkenrode*, **6,** 91 (1956)
154. Wickersheim, K. A., *J. chem. Phys.*, **29,** 1197
155. Phillips, B. A. and Busing, W. R., *J. phys. Chem.*, **61,** 502
156. Simon, A. and Schmidt, W., *Z. Elektrochemie*, **64,** 737 (1960)
157. Mara, R. T. and Sutherland, G. B. B. M., *J. opt. Soc. Am.*, **43,** 1100 (1953)
158. Jones, L. H., *J. chem. Phys.*, **22,** 127 (1954)
159. Decius, J. C., *J. chem. Phys.*, **22,** 1941 (1954)
160. Wiberley, S. E., Sprague, J. W. and Campbell, J. E., *Analyt. Chem.*, **29,** 210 (1957)
161. Duval, C. and Lecomte, J., *Bull. Soc. chim. Fr.*, 1057 (1947)
162. Sun, M.-S., *Am. Miner.*, **48,** 655 (1963)
163. Duval, C. and Lecomte, J., *J. opt. Soc. Am.*, **44,** 261 (1954)

164. Miller, J. G., *J. phys. Chem.*, **65**, 800 (1961)
165. Detoni, S. and Hadzi, D. *J. Chim. phys.*, **53** (9), 760 (1956)
166. Tarte, P., *Spectrochim. Acta*, **13**, 107 (1958)
167. Stammreich, H., Bassi, D., Sala, O. and Siebert, H., *Spectrochim. Acta*, **13**, 192 (1958)
168. Tuddenham, W. M. and Lyon, R. J. P., *Analyt. Chem.*, **31**, 377 (1959)
169. Tarte, P. and Collongues, R., *Ann. Chim.*, **9**, 135 (1964)
170. Barrow, G. M., *J. chem. Phys.*, **21**, 219 (1953)
171. Goulden, J. D. S., *Spectrochim. Acta*, **15**, 657 (1959)
172. Plyusnina, I. I., *Geochemistry*, No. 2, p. 174 (1963)
173. Petch, H. E., Sheppard, N. and Megaw, H. D., *Acta Crystallogr.*, **9**, 29 (1956)
174. Martius, U. M. and Kirby, E. M., *8th Ottawa Symp. Appl. Spectrosc.*
175. Haas, M. and Sutherland, G. B. B. M., *Proc. roy. Soc.*, Ser. A, **236**, 427 (1956)
176. Duval, C. and Lecomte, J., *Comptes Rendus hebd. Séanc. Acad. Sci., Paris*, **232**, 2306 (1951)
177. Pepperhof, W., *Arch. EisenhüttWes.*, **29**, 153 (1958)
178. Bunting, E. N. and Van-Valkenburg, A., *Am. Miner.*, **43**, 102 (1958)
179. Berry, E. E., *Spectrochim. Acta*, **24A**, 1727 (1968)
180. Hilsum, C., *J. opt. Soc. Am.*, **45**, 771 (1955)
181. Duval, C. and Lecomte, J., *Comptes Rendus hebd. Séanc. Acad. Sci., Paris*, **239**, 249 (1954)
182. Glemser, O. and Hartert, E., *Z. anorg. allg. Chem.*, **283**, 111 (1956)
183. Wickersheim, K. A., Lefever, R. A. and Hanking, B. M., *J. chem. Phys.*, **32**, 271 (1960)
184. Stewart, J. E., *Infrared Spectroscopy*, New York, Dekker (1970)
185. Glasner, A. and Reisfeld, R., *J. chem. Phys.*, **32**, 956 (1960)
186. Ramdas, A. K., *Proc. Indian Acad. Sci.*, Sec. A, **37** (3), 451 (1953)
187. Hafele, H. G., *Z. Phys.*, **160**, 420 (1960)
188. Tsuboi, M., *Bull. chem. Soc. Japan*, **23** (3), 83 (1950)
189. Lecomte, J. and Freymann, R., *Bull. Soc. chim. Fr.*, **8**, 622 (1941)
190. Goubeau, J. and Hummel, D., *Z. phys. Chem.*, **20**, 15 (1959)
191. Von Muller-Hesse, H. and Schwiete, H. E., *Sprechsaal Keram. Glas Email*, **93**, 147 (1960)
192. Von Muller-Hesse, H., Planz, E. and Schwiete, H. E., *Sprechsaal Keram. Glas Email*, **93**, 159 (1960)
193. Petch, H. E. and Megaw, H. D., *J. opt. Soc. Am.*, **44**, 744 (1954)
194. Matossi, F. and Brix, H., *Z. Phys.*, **92**, 303 (1934)
195. Matossi, F. and Kindler, H., *Z. Phys.*, **92**, 312 (1934)
196. Corbridge, D. E. C. and Lowe, E. J., *J. chem. Soc.*, 4555 (1954)
197. Hartert, E. and Glemser, O., *Z. Elektrochemie*, **60** (7), 746 (1956)
198. Florinskaya, V. A. and Pechenkina, R. S., *Izv. Akad. Nauk. SSSR Ser. Fiz.*, **17** (5), 649 (1953)
199. Simeral, W. G., *Dissert. Abstr.*, **13**, 840 (1953)
200. Mitsuishi, A., Yoshinaga, H. and Fujita, S., *J. phys. Soc. Japan*, **13**, 1235 (1958)
201. Moenke, H., V.E.B. *Carl Zeiss Jena Nachrichten*, p. 49 (Sept. 1961)
202. Moenke, H., V.E.B. *Carl Zeiss Jena Nachrichten*, p. 82 (Sept. 1961)
203. Sutherland, G. B. B. M., Blackwell, D. E. and Simeral, W. G., *Nature*, **174**, 901 (1954)
204. Stubican, V. and Gunthard, H., *Nature*, **179**, 542 (1957)
205. Hacskaylo, M., *Analyt. Chem.*, **26**, 1410 (1954)
206. Ramanthan, K. G., *Proc. Indian Acad. Sci.*, **26A** (6), 469 (1947)
207. Mitsuishi, A., Yoshinga, H. and Fujita, S., *J. phys. Soc. Japan*, **13**, 1236 (1958)
208. Ramdas, A. K., *Proc. Indian Acad. Sci.*, **39A** (2), 81 (1954)
209. Hidalgo, A. and Serratosa, J. M., *Nouvo cimento Suppl.*, Series 10, **2**, 742 (1955)
210. Sevchenko, N. A. and Florinskaya, V. A., *Dokl. Akad. Nauk SSSR*, **109**, 1115 (1956)
211. Keller, W. D. and Pickett, E. E., *Am. J. Sci.*, **252**, 87 (1954)
212. Duval, R. and Lecomte, J., *Comptes Rendus hebd. Séanc. Acad. Sci., Paris*, **213**, 999 (1941)
213. Cabannes-Ott, C., *Comptes Rendus hebd. Séanc. Acad. Sci., Paris*, **242**, 2825 (1956)
214. Rocchiccioli, C., *Comptes Rendus hebd. Séanc. Acad. Sci., Paris*, **242**, 2922 (1956)
215. Bergmann, E. D., Littauer, U. Z. and Pinchas, S., *J. chem. Soc.*, 847 (1952)
216. Drouard, E., *Comptes Rendus hebd. Séanc. Acad. Sci., Paris*, **240**, 1700 (1955)
217. Gebbie, H. A. and Cannon, C. G., *J. opt. Soc. Am.*, **42** (7), 277 (1952)
218. Cohn, H., *J. chem. Soc.*, 4282 (1952)
219. Falk, M. and Giguere, P. A., *Can. J. Chem.*, **36**, 1680 (1958)
220. Gatehouse, B. M., Livingstone, S. E. and Nyholm, R. S., *J. chem. Soc.* (*Lond.*), 3137 (1958)
221. Montova, J. W. and Baur, G. S., *Am. Miner.*, **48**, 1227 (1963)
222. Rocchiccioli, C., *Comptes Rendus hebd. Séanc. Acad. Sci., Paris*, **247**, 1108 (1958)
223. Prigent, J., *Comptes Rendus hebd. Séanc. Acad. Sci., Paris*, **247**, 1737 (1958)
224. Ramdas A. K., *Proc. Indian Acad. Sci.*, **36A**, 55 (19 52)

225. Babushkin, A. A., Yukhnevich, G. V. *et al.*, *Zh. neorg. Khim.*, **4,** 823 (1959)
226. Akhmanova, M. V., Karyakin, A. V. and Yukhnevich, G. V., *Geochem.*, **6**, 596 (1963)
227. Stephens, J.D., Presented to 1963 meeting of the Geological Society of America in New York City
228. Tuddenham, W. M. and Lyon, R. J. P., *Analyt. Chem.*, **32** (12), 1630 (1960)
229. Dasgupta, S. and Sinha, A. P. B., *Trans. Faraday Soc.*, **53,** 909 (1957)
230. Tsuboi, M., *J. Am. chem. Soc.*, **79,** 1351 (1957)
231. Last, J. T., *Phys. Rev.*, **105** (6), 1740 (1957)
232. Dupuis, T., *Comptes Rendus hebd. Séanc. Acad. Sci., Paris*, **246,** 3332 (1958)
233. Evans, J. C. and Bernstein, H. J., *Can. J. Chem.*, **33,** 1270 (1955)
234. Runnells, D. D., *Am. Miner.*, **49,** (1 and 2), 158 (1964)
235. DeKimpe, C. and Gastuche, M. C., *Am. Miner.*, **49,** (1 and 2), 1 (1964)
236. Laforge-Kantzer, D., *Annls. Chimie*, **5,** 819 (1950)
237. Duval, C., Lecomte, J. and Morandat, J., *Bull. Soc. chim. Fr.*, 745 (Sept–Oct 1951)
238. Hadni, A. and Michel, R. L., *Comptes Rendus hebd. Séanc. Acad. Sci., Paris*, **244,** 2602 (1957)
239. Louisfert, J., *Comptes Rendus hebd. Séanc. Acad. Sci., Paris*, **233,** 381 (1951)
240. Martinez, E., *Trans. Soc. mining Engrs.*, **226,** 424 (1963)
241. Waddington, T. C., *J. chem. Soc.*, 4340 (1958)
242. Gatehouse, B. M. and Comyns, A. E., *J. chem. Soc.*, 3965 (1958)
243. Glemser, O. and Hartert, E., *Naturwissenschaften*, **40,** 552 (1953)
244. Glemser, O. and Hartert, E., *Naturwissenschaften*, **42,** 534 (1955)
245. Arkhipenko, D. K., *J. struct. Chem.*, **4** (2), 180 (1963)
246. Mikheeva, Y. T. and Mal'tseva, N. N., *J. struct. Chem.*, **4** (5), 643 (1963)
247. Lazarev, A. N. and Tenisheya, T. F., *J. struct. Chem.*, **4** (5), 647 (1963)
248. Batsanova, L. P., Grigor'eva, G. N. and Batsanova, S. S., *J. struct. Chem.*, **5** (1), 33 (1963)
249. Chalmers, R. A., Farmer, V. C. *et al. Mineralog. Mag.*, **33,** 821 (1964)
250. Strens, R. G. J., *Mineralog. Mag.*, **33,** 868 (1964)
251. Vedder, W. *Am. Miner.*, **49,** 736 (1964)
252. Erd, R. C., White, D. E. *et al.*, *Am. Miner.*, **49,** 831 (1964)
253. Haccuria, M., *Bull. Soc. chim. Belges*, **62,** 428 (1953)
254. Bovey, L. F. H., *J. opt. Soc. Am.*, **41,** 836 (1951)
255. Terada, M. and Tsuboi, M., *Bull. chem. Soc. Japan*, **37,** 1080 (1964)
256. Mutschin, A. and Maennchen, K., *Z. analyt. Chem.*, **156** (4), 241 (1957)
257. Adler, H. H. and Kerr, P. F., *Am. Miner.*, **49,** 1002 (1964)
258. Wise, W. S. and Eugster, H. P., *Am. Miner.*, **49,** 1031 (1964)
259. Price, W. C. and Wilkinson, G. R., *J. phys. Chem. Solids*, **18** (1), 74–76 (1961)
260. Van-Valkenberg, A. and Rynders, G. F., *Am. Miner.*, **43,** 1195 (1958)
261. Roy, D. M., *Mineralog. Mag.*, **31,** 187 (1956)
262. Peacock, R. D. and Sharp, D. W. A., *J. chem. Soc.*, 2762 (1959)
263. Reitzel, J., *J. chem. Phys.*, **23,** 2407 (1955)
264. Roy, R. and Francis, E. E., *Am. Miner.*, **38,** 725 (1953)
265. Pistorius, C. W. F. T., *J. inorg. nucl. Chem.*, **15,** 187 (1960)
266. Angino, E. E., *Nature*, **204,** 569 (1964)
267. Waddington, T. C., *J. chem. Soc.*, 2499 (1959)
268. Newman, R. and Halford, R. S., *J. chem. Phys.*, **18,** 1291 (1950)
269. Fripiat, J. J., Chaussidon, J. and Touillaux, R., *J. phys. Chem.*, **64,** 1234 (1960)
270. Roy, R., Roy, D. M. and Francis, E. E., *J. Am. ceram. Soc.*, **38,** 198 (1955)
271. Giguere, P. A. and Falk, M., *Spectrochim. Acta*, **16** pl. (1960)
272. Klimov, V. V., *Chem. Abstr.*, **55,** 9036 g (1961)
273. Vratny, F., Tsai, M. and Gugliotta, F., *Nature*, **188,** 484 (1960)
274. Tai, H., *Dissert. Abstr.*, **19,** 2734 (1959)
275. Fisher, R. B. and Ring, C. E., *Analyt. Chem.*, **29,** 431 (1957)
276. Ginzburg, I. V. and Yukhnevich, G. V., *Geochemistry*, **1,** 31 (1962)
277. Meixner, H. and Moenke, H., *Kali Steinsalz*, **3,** 228 (1961)
278. Brunner, G., Wondratschek, H. and Laves, F., *Fortschr. Miner.*, **38,** 125 (1960)
279. White, W. B. and Roy, R., *Am. Miner.*, **49,** 1670 (1964)
280. Grisape, D. A. and White, W. B., *Am. Miner.*, **49,** 1184 (1964)
281. Duke, D. A. and Stephens, J. D. *Am. Miner.*, **49,** 1388 (1964)
282. Thompson, C. S. and Wadsworth, M. E. *Am. Miner.*, **42,** 334 (1957)
283. Hafele, H. G., *Z. Phys.*, **148,** 262 (1957)
284. Steger, E. and Leukroth, G., *Z. anorg. Allg. Chem.*, **303,** 169 (1960)
285. Siebert, H., *Z. anorg. Allg. Chem.*, **301,** 161 (1959)

286. Soda, R., *Bull. chem. Soc. Japan*, **34**, 1491 (1961)
287. Stekhanov, A. I., *Izv. Akad. Nauk. SSSR*, Ser. F12, **22**, 1109 (1958)
288. Winand, L. and Duyckaerts, G., *Bull. Soc. chim. Belges*, **71**, 142 (1962)
289. Lyon, W. and Kinsey, E. L., *Phys. Rev.*, **61**, 482 (1942)
290. Lecomte, J. and Duval, C., *J. Chim. phys.* **50**, C53, (1953)
291. Wolff, R. G., *Dissert. Abstr.*, XII No. 5, 1652 (1961)
292. Sato, T., *J. appl. Chem.*, **11** (6), 207 (1961)
293. Sato, T., *J. appl. Chem.*, **12** (1), 9 (1962)
294. Austin, I. G., Gebbie, H. A. and Morin, I., *Proc. phys. Soc.*, Pt. 6 (i), No. 505, **78**, 1260 (1961)
295. Bishui, B. M. and Prasad, J., *Bull. cent. Glass Ceram. Res. Inst., India*, **7**, 97 (1960)
296. Koltermann, M., *Ber. dt. keram. Ges.*, **38**, 543 (1961)
297. Corbridge, D. E. C. and Lowe, E. J., *Analyt. Chem.*, **27**, 1383 (1955)
298. Lyon, R. J. P., *Nature*, **196**, 266 (1962)
299. Sclar, C. B., Carrison, L. C. and Schwartz, C. M., *Science*, **138**, 525 (1962)
300. Saksena, B. D., *Proc. Indian Acad. Sci.*, p. 93 (July 1940)
301. Saksena, B. D., *Trans. Faraday Soc.*, **57**, 242 (1961)
302. Rocchiccioli, C., *Annls. Chimie*, **5**, 999 (1960)
303. Lefevre, R. J. W., Or, W. T. *et al.*, *Aust. J. Chem.*, **10** (3), 361 (1957)
304. Lippincott, E. R., Van Valkenberg, A. *et al.*, *J. Res. natn Bur. Stand.*, **61** (1), 61 (July 1958)
305. Pudovkina, E. A., Pudovkina, Z. B. and Solntzeva, L. S., *Mineralog. Sb., L'vov.*, **15**, 120 (1961)
306. Archipenko, D. K., *Geologiya Geofiz*, **3**, 115 (1962)
307. Mutschin, A. and Maennchen, K., *Z. anal. Chim.*, **160** (2), 81 (1958)
308. Chapman, A. C. and Thirwell, L. E., *Spectrochim. Acta*, **20**, 937 (1964)
309. Sathianandan, K., McCory, L. D. and Margrave, J. L., *Spectrochim. Acta*, **20**, 957 (1964)
310. D'Or, L., Haccuria, M. and Machiroux, R., *C.r. 27th congr. inter. chim. ind.*, **3**, 150 (1954, pub. 1955)
311. Touillaux, R., Fripiat, J. J. and Touissant, F., *Trans. 7th Internat. Congr. Soil Sci.*, **4**, 460 (1960)
312. Manghnani, M. H. and Hower, J., *Am. Miner.*, **49**, 1631 (1964)
313. Adler, H. H. and Kerr, P. F., *Am. Miner.*, **48**, 839 (1963)
314. Ta, Y., *Comptes Rendus hebd. Séanc. Acad. Sci., Paris*, **211** (20), 467 (1940)
315. Kalbus, G. E., *Doct. Dissert.*, *Pub. No.* 24,300, *Ann. Arbor*, Michigan, Univ. Microfilms (1957)
316. Fripiat, J. J., *Bull. Grpe fr. Argiles*, **12**, 25 (1960)
317. Serratosa, J. M. and Bradley, W. F., *Nature*, **181**, 111 (1958)
318. Frederickson, L. D., *Analyt. Chem.*, **26**, 1883 (1954)
319. Kuhn, L. and Lippincott, E. R., *J. Am. chem. Soc.*, **78**, 1820 (1956)
320. Lucchesi, P. J. and Glasson, W. A., *J. Am. chem. Soc.*, **78**, 1347 (1956)
321. Cabannes-Ott, C., *Annls Chimie*, **5**, 905 (1960)
322. Caspers, H. H., Buchanan, R. A. and Marlin, H. R., *J. chem. Phys.*, **41**, 94 (1964)
323. Busey, R. H. and Keller, O. L., *J. chem. Phys.*, **41**, 215 (1964)
324. Jones, L. H. and Smith, J. M., *J. chem. Phys.*, **41**, 2507 (1964)
325. Plyusnina, I. I., *Geochem. Inter.*, **1**, 13 (1964)
326. Stubican, V. S., *Mineral Ind.*, **33** (2) pl. (1963)
327. Farmer, V. C., *Mineralog. Mag.*, **31**, 829 (1958)
328. Keller, W. D., *Bull. geol. Soc. Am.*, **59**, 1332 (1948)
329. Sevchenko, N. A. and Florinskaya, V. A., *Sov. Phys.*, **1** (4), 508 (1956)
330. Ramdas, A. K., *Proc. Indian Acad. Sci.*, Sec. A, **35**, 249 (1952)
331. Dachille, F. and Roy, R., *Z. Krist.*, **111**, 462 (1959)
332. Lecomte, J., *Cah. Phys.*, Nos. 17–21, pl-26 (1943–44)
333. Nahin, P. G., Merrill, W. C. *et al.*, *J. Pet. Techn. Am. Inst. Min. Met. Eng. Petroleum Trans.*, **192**, 151 (1951)
334. Tarte, P., *Silic. Ind.*, **24**, 7 (1959)
335. Tarte, P., *Silic. Ind.*, **25**, 171 (1960)
336. Cavell, R. G. and Clark, H. C., *Inorg. Chem.*, **3**, 1789 (1964)
337. Dupuis, T., *Mikrochim. Acta.*, **4**, 851 (1955)
338. Beutelspacher, H., *6th Int. Congr. Soil Sci., Paris*, p. 329 (1965)
339. Datta, R. K. and Roy, R., *Nature*, **191**, 169 (1961)
340. Hafner, St. and Laves, F., *Z. Kristallogr.*, **109**, 204 (1957)
341. Bassett, W. A., *Bull. geol. Soc. Am.*, **71**, 449 (1960)
342. Hexter, R. M., *Spectrochim. Acta*, **10**, 291 (1958)
343. Jones, L. H., *Spectrochim. Acta*, **10**, 395 (1958)
344. Duval, C., *IIIrd Congr. Int. Reactions a l'etat Solide, Madrid*, (1956)
345. Kalousek, G. L. and Roy, R., *J. Am. ceram. Soc.*, **40**, 236 (1957)

346. Kolesova, V. A., *Z. Exp. teoret. Fiz.*, **26** (1), 124 (1954)
347. Louisfert, J., *J. phys. Rad.*, **8,** 21 (1947)
348. Duval, C. and Lecomte, J., *Bull. Soc. chim. Fr.*, **5,** 713 (1941)
349. Rocchiccioli, C., *Comptes Rendus hebd. Séanc. Acad. Sci., Paris*, **250,** 1232 (1960)
350. Rocchiccioli, C., *Comptes Rendus hebd. Séanc. Acad. Sci., Paris*, **249,** 236 (1959)
351. Adams, R. V. and Douglas, R. W., *J. Soc. glass Tech.*, **43,** 147 (1959)
352. Hodgson, A. A. and Freeman, A. G., *S. Afr. Min. Mag.*, **35** (269), 5 (1965)
353. Ferraro, J. R. and Walker, A., *J. chem. Phys.*, **42** (4), 1273 (1965)
354. Ferraro, J. R. and Walker, A., *J. chem. Phys.*, **42** (4), 1278 (1965)
355. Koningstein, J. A., *J. chem. Phys.*, **42** (4), 1423 (1965)
356. Von Hauschild, U., *Z. anorg. Allg. Chem.*, **324,** 15 (1963)
357. Dupuis, T., *Mikrochim. Acta*, **2–4,** 228 (1964)
358. Bethell, D. E. and Sheppard, N., *Trans. Faraday Soc.*, **51,** 9 (1955)
359. Blinc, R., Maricic, S. and Pintar, M., *Croat. chem. Acta*, **32,** 67 (1960)
360. Schwarcz, H. P. and Speelman, E. L., *Am. Miner.*, **50,** 656 (1965)
361. Low, M. J. D., *Nature*, **208,** 1089 (1965)
362. Iberg, R., *Helv. chem. Acta*, **40,** 102 (1957)
363. Krilov, G. M. and Preyev, Ya. M., *Dokl. Akad. Nauk. Uxbek SSSR.*, No. 1, p. 25 (1959)
364. Bokii, G. B. and Plyusnina, E. E., *Nauk. Dokl. vysh. shkoly geol. geogr.*, **3,** 116 (1958)
365. Hartert, E., *Naturwissenschaften*, **43,** 275 (1956)
366. Laves, F. and Hafner, St., *Z. Kristallogr.*, **108,** 52 (1956)
367. Shimida, K. and Komaki, T., *J. chem. Soc. Japan*, **72** (8), 938 (1951)
368. Hidalgo, A. and Serratosa, J. M., *Anales adafo fis. veg.* (*Madrid*), **14,** 269 (1955)
369. Khanna, R. K., Lingscheid, J. and Decius, J. C., *Spectrochim. Acta*, **20,** 1106 (1964)
370. Farmer, V. C. and Russell, J. D., *Spectrochim. Acta*, **20,** 1149 (1964)
371. Ross, S. D. and Goldsmith, J. A., *Spectrochim. Acta*, **20,** 781 (1964)
372. Winchell, A. N. and Winchell, H., 'Elements of Optical Mineralogy, Part II', in *Descriptions of Minerals*, 4th ed., Wiley & Sons (1956)
373. Akhmanova, M. V. and Leonova, L. L. *Geokhim.*, **5,** 401 (1961)
374. Adler, H. H. *Am. Miner.*, **50,** 1553 (1965)
375. Fisher, D. J., *Am. Miner.*, **50,** 1647 (1965)
376. Arlidge, E. Z. and Farmer, V.C., *J. appl. Chem.*, **13,** 17 (1963)
377. Goldsmith, J. A. and Ross, S. D., *Spectrochim. Acta*, **22,** 1069 (1966)
378. Carpenter, A. B., Chalmers, R. A. *et al.*, *Am. Miner.*, **51,** 56 (1966)
379. Newkirk, H. W., Smith, D. K. and Kahn, J. S., *Am. Miner.*, **51,** 141 (1966)
380. Kostova, G. N. and Batsanova, P. R., *Zh. prikl. Spektrosk.*, **4,** 75 (1966)
381. Mumpton, F. A., Jaffa, H. W. and Thompson, C. S., *Am. Miner.*, **50,** 1893 (1965)
382. McDevitt, N. T. and Davidson, A. D., *J. opt. Soc. Am.*, **56** (5), 636 (1966)
383. Deer, W. A., Howie, R. A. and Zussman, J., *Rock Forming Minerals*, Vols 1–5, Longmans, London (1962–63)
384. Kohls, D. W. and Rodda, J. L., *Am. Miner.*, **51,** 677 (1966)
385. Ryall, W. R. and Threadgold, I. M. *Am. Miner.*, **51,** 754 (1966)
386. Gulbrandsen, R. A., Kramer, J. R. *et al.*, *Am. Miner.*, **51,** 819 (1966)
387. Semenov, E. I. *et al.*, *Meddr Gronland*, **181,** 1 (1965)
388. Brehat, F., Evrard, O. *et al.*, *Comptes Rendus hebd. Séanc. Acad. Sci., Paris*, No. 20, **263B,** 1112 (1966)
389. Pecile, C., *Inorg. Chem.*, **5** (2), 210 (1966)
390. Petru, F. and Muck, A., *Z. Chem.*, **6,** 386 (1966)
391. Schwarzmann, E. and Marshmann, H., *Z. Naturf.*, **21b,** 924 (1966)
392. Kharetonov, Yo. Ya., Tverdokhlebov, V. N. and Molodin, A. K., *Zh. neorg. Khem.*, **11,** 2608 (1966)
393. Ostrovskaya, I. V., Pertsev, N. N. and Nikitina, I. V., *Zap. vses. miner. Obshch.*, 193 (1966)
394. Palenko, A. S., Orlova, L. P. and Achmanova, M. V., *Trudy miner. muzeya Akad. Nauk SSSR*, **16,** 166 (1965)
395. Pertsev, N. N., Ostrovskaya, E. V. and Nikitina, I. V., *Z. vses. miner. Obshch.*, **94,** 178 (1965)
396. Muller, A. and Krebs, B., *Z. Naturf.*, **21b,** 3 (1966)
397. Laves, F. and Hafner, St., *Z. Kristallogr.*, **108,** 52 (1956)
398. Stubican, V., and Roy, R., *Z. Kristallogr.*, **115,** 200 (1961)
399. Akhmanova, M. V. and Orlova, L. P., *Geochem. Internat.*, **3** (3), 444 (1966)
400. Angino, E. E., *Am. Miner.*, **52,** 137 (1967)
401. Bunch, T. E. and Cohen, A. J., *Am. Miner.*, **52,** 244 (1967)
402. Wolff, R. G., *Am. Miner.*, **50,** 240 (1965)

403. Adler, H. H. and Kerr, P. F., *Am. Miner.*, **50**, 132 (1965)
404. Harada, K. and Tomita, K., *Am. Miner.*, **52**, 1438 (1967)
405. Flanigen, E. M., Breck, D. W. *et al.*, *Am. Miner.*, **52**, 744 (1967)
406. Liese, H. C., *Am. Miner.*, **52**, 877 (1967)
407. Liese, H. C., *Am. Miner.*, **52**, 1198 (1967)
408. Von-Rawden, H. V. R. and Dicks, L. W. R., *Am. Miner.*, **52**, 1067 (1967)
409. Hayashi, H. and Oinuma, K., *Am. Miner.*, **52**, 1206 (1967)
410. Batsanova, S. S. and Derbenyova, S. S., *Optics Spectrosc.*, **17** (1), 78 (1964)
411. Baran, E. J. and Aymonino, P. J., *Spectrochim. Acta*, **24A**, 288 (1968)
412. Baran, E. J. and Aymonino, P. J., *Spectrochim. Acta*, **24A**, 291 (1968)
413. Steger, E. and Kabner, B., *Spectrochim. Acta*, **24A**, 447 (1968)
414. Ohwada, K., *Spectrochim. Acta*, **24A**, 595 (1968)
415. Hezel, A. and Ross, S. D., *Spectrochim. Acta*, **22**, 1949 (1966)
416. McDevitt, N. T. and Baun, W. L., *Spectrochim. Acta*, **20**, 799 (1964)
417. Laperches, J. P. and Tarte, P., *Spectrochim. Acta*, **22**, 1201 (1966)
418. Ahlijah, G. E. B. Y. and Mooney, E. F., *Spectrochim. Acta*, **22**, 547 (1966)
419. Khanna, R. K. and Lippincott, E. R., *Spectrochim. Acta*, **24A**, 905 (1968)
420. Brophy, G. P. and Nash, J. T., *Am. Miner.*, **53**, 445 (1968)
421. Hunt, C. M., *4th Internat. Symp. Chem. of Cement, Washington, D.C.*, Pap. III, S8 (1960)
422. Haas, C. and Ketelaar, J. A. A., *Physica*, **XXII**, 1286
423. Clark, G. M. and Doyle, W. P., *Spectrochim. Acta*, **22**, 1441 (1966)
424. Kirov, G. N. and Poulieff, C. N., *Mineralog. Mag.*, **32**, 1003 (1968)
425. Bhatnagar, V. M., *Experentia*, **23/1, 15–1,** 10 (1967)
426. Tramer, A., *J. chim. Phys.*, **59**, 232 (1962)
427. Pecile, C., Giacometti, G. and Turco, A., *Lincei-Rend. Sc. fis. mat. e. nat.*, **28**, 189 (1960)
428. Buijs, K. and Schutte, C. J. H., *Spectrochim. Acta*, **17**, 917 (1961)
429. Buijs, K. and Schutte, C. J. H., *Nature*, **192**, 351 (1961)
430. Farrell, E. F. and Newnham, R. E., *Am. Miner.*, **52**, 380 (1967)
431. Omori, K. and Kerr, P. F., *Geol. soc. am. Bull.*, **74**, 709 (1963)
432. Adler, H. H., *Am. Miner.*, **53**, 1741 (1968)
433. Datta, R. K. and Roy, R., *Am. Miner.*, **53**, 1456 (1968)
434. Ryall, W. R. and Threadgold, I. M., *Am. Miner.*, **53**, 1614 (1968)
435. Steger, E. and Schmidt, W., *Ber. bunsenges. phy. Chem.*, **68** (1), 102 (1964)
436. Decius, J. C., Coker, E. H. and Brenna, G. L., *Spectrochim. Acta*, **19**, 1281 (1963)
437. Tarte, P. and Nizet, G., *Spectrochim. Acta*, **20**, 503 (1964)
438. Wolff, R. G., *Am. Miner.*, **50**, 240 (1965)
439. Campbell, J. A., *Spectrochim. Acta*, **21**, 851 (1965)
440. Paques-Ledent, M. Th. and Tarte, P., *Spectrochim. Acta*, **25A**, 1115 (1969)
441. Ahlijah, G. E. B. Y. and Mooney, E. F., *Spectrochim. Acta*, **25A**, 619 (1969)
442. De Abeledo, M. A. J. and De Benyacar, M. A. R., *Am. Miner.*, **53**, 2077 (1968)
443. Muller, O., White, W. B. and Roy, R., *Spectrochim. Acta*, **25A**, 1491 (1969)
444. Liese, H. C., *Am. Miner.*, **48**, 980 (1963)
445. Vedder, W. and Wilkins, R. W. T., *Am. Miner.*, **54**, 483 (1969)
446. Zdenek, J., Povondra, P. and Slanksy, E., *Am. Miner.*, **54**, 1 (1969)
447. Guerchais, J. E., Leroy, M. J. and Rohmer, R., *Comptes Rendus hebd. Séanc. Acad. Sci., Paris*, **261** (Gp 8), 3628 (1965)
448. Jaulmes, S., *Rev. chim. Min.*, **2**, 147 (1965)
449. Caro, P. E., Sawyer, J. O. and Eyring, L. *Spectrochim. Acta*, **28A**, 1167 (1972)
450. Wickersheim, K. A. and Lefever, R. A., *J. Am. chem. Soc.*, **83**, 1147 (1961)
451. Ariya, S. M. and Golomolzina, M. V., *Sov. Phys.—solid state*, **4** (10), 2142 (1963)
452. Valarelli, J. V., Perrier, M. and Vincentini, G., *An. acad. Brasil Cienc.*, **40** (3), 289 (1968)
453. Gevorkyan, S. V. and Povarennykn, A. S., *Dop. akad. nauk, Ukr. RSR*, Ser. B, **31** (4), 298 (1969)
454. Taylor, R. G., Morgan, W. R. and Phillips, D. N., *Mineralog. Mag.*, **37** (289), 624 (1970)
455. Ehlmann, A. J. and Mitchell, R. S., *Am. Miner.*, **55** pl. (1970)
456. DeWall, S. A., *Am. Miner.*, **55**, 31 (1970)
457. Sclar, C. B., Carrison, L. C. and Schwartz, C. M., *Science*, **138**, 525 (1962)
458. Tien, P. L. and Waugh, T. C., *Am. Miner.*, **54**, 1355 (1969)
459. Hariya, Y., Dollase, W. A. and Kennedy, G. C., *Am. Miner.*, **54**, 1419 (1969)
460. Langer, K. and Schreyer, W., *Am. Miner.*, **54**, 1442 (1969)
461. Rooney, R. P., Jones, B. F. and Neal, J. T., *Am. Miner.*, **54**, 1034 (1969)
462. Forneris, R., *Am. Miner.*, **54**, 1062 (1969)

463. Darrie, R. G., Doyle, W. P. and Kirkpatrick, I., *J. inorg. nucl. Chem.*, **29,** 979 (1967)
464. Simon, I. and McMahon, W. O., *J. chem. Phys.*, **21** (1), 23 (1953)
465. Dune, R. A. and Rosevear, P. E., *Inv. Rep. 85 of the Div. of Min. Chem.*, S.I.R.O. Sydney Laboratory (July 1970)
466. Campbell, J. A., Ryan, D. P. and Simpson, L. M., *Spectrochim. Acta,* **26A,** 2351 (1970)
467. Rutstein, M. S. and White, W. B., *Am. Miner.*, **56,** 877 (1971)
468. Omori, K., *Am. Miner.*, **56,** 841 (1971)
469. Kodama, H. and Ginuma, K., *Proc. 11th Nat. Congr.*, p. 236
470. Wehrenberg, J. P., *Am. Miner.*, **56,** 1639 (1971)
471. von Schwarzmann, E., *Z. anorg. Chem.*, **317,** 176 (1962)
472. Ross, S. D., *Spectrochim. Acta,* **27A,** 1837 (1971)
473. Greenwood, N. N., Ross, E. J. F. and Straughan, B. P., *Index of Vibrational Spectra of Inorganic and Organometallic Compounds*, London, Butterworths (1972)
474. Ferraro, J. R., *J. mol. Spectr.*, **4,** 99 (1960)
475. Nakamoto, K., *J. Am. chem. Soc.*, **79,** 4904 (1957)
476. Ross, S. D., *Spectrochim. Acta,* **18,** 225 (1962)
477. Karr Jr., C., *Prog. Nucl. Energy Ser. IX Analyt. Chem.* **11,** 111 (1972)
478. Jones, L. H. and Penneman, R. A., *J. chem. Phys.*, **21,** 542 (1953)
479. Stubican, V. and Roy, R., *J. phys. Chem.*, **65,** 1348 (1961)
480. Keller, W. D. and Pickett, E. E., *Am. J. Sci.*, **248,** 264 (1950)
481. Luce, R. W., *U.S. Geolog. Surv. Prof.*, Pap. 750B, p. 199 (1971)
482. Billhardt, H. W., *Am. Miner.*, **54,** 511 (1969)
483. Clark Jr., S. P., *Am. Miner.*, **42,** 732 (1957)
484. Wollast, R., Mackenzie, F. T. and Bricker, O. P., *Am. Miner.*, **53,** 1651 (1968)
485. Hayashi, H. and Oinuma, K., *Am. Miner.*, **50,** 476 (1965)
486. Angino, E. E., *Kansas Geolog. Surv. Bull.*, **194** (1), 9 (1969)
487. Garcia, S. G., *Annls fis. Quim.*, p. 39 (1960)
488. Ferraro, J. R., 'Interpretation of the Infrared Spectra of Inorganic Molecules', in *Prog. Infrared Spectrosc., Vol. 1, Proc. Vth Ann. Infrared Spectrosc. Conf.* (*Aug. 61*) (ed. Szymanski), Plenum Press, New York (1962)
489. Ferraro, J. R., *J. chem. Educ.*, **38,** (4), 201 (1961)
490. Brydon, J. E. and Kodama, H., *Am. Miner.*, **51,** 875 (1966)
491. Jones, L. H., *Spectrochim. Acta,* **15,** 409 (1959)
492. Ketelaar, J. A. A., Schutte, C. J. H. and Schram, B. L. S., *Spectrochim. Acta,* **13,** 336 (1958)
493. Newnham, R. E., *Mineralog. Mag.*, **32,** 683 (1961)
494. Moenke, H., *Infrared Phys.*, **2,** 111 (1962)
495. Hafner, S., *Z. Kristallogr.*, **115,** 331 (1963)
496. Stubican, V., *Mineralog. Mag.*, **32,** 38 (1959)
497. Stephens, J. D. and Tuddenham, W. M., Paper presented at the 1963 Meeting of A.I.M.E. at Dallas, Texas
498. Antikainen, P. J., *Suom. Kem.*, **31,** 223 (1958)
499. Dachille, F. and Roy, R., *Am. J. Sci.*, **258,** 225 (1960)
500. Lazarev, A. N., *Vibrational Spectra and Structure of Silicates*, New York/London, Consultants Bureau (1972)
501. Wilkins, R. W. T. and Ito, J., *Am. Miner.*, **52,** 1649 (1967)
502. Oinuma, K. and Hayashi, H., *Am. Miner.*, **50,** 1213 (1965)
503. Stubican, V. and Roy, R., *Abstracts of the 137th Meeting of the Am. Chem. Soc.*, Sect. 1, p. 20 (April 1960)
504. Tettenhorst, R., *Am. Miner.*, **47,** 769 (1962)
505. White, W. B., *Am. Miner.*, **56,** 46 (1971)
506. Baron, G., Caillere, S., Lagrange, R. and Pobeguin, Th., *Bull. Soc. fr. Miner. Cristallogr.*, **82,** 150 (1959)
507. Greenwood, N. N., Ross, E. J. F. and Straughan, B. P., *Index of Vibrational Spectra of Inorganic and Organometallic Compounds*, London, Butterworths (1972)
508. Taylor, D. G., Nenadic, C. M. and Crable, J. V., *Am. Ind. Hyg. Ass. J.*, **31** (1), 100 (1970)
509. Levitt, S. R. and Condrate, Sr., R. A., *Am. Miner.*, **55,** p. 1562 (1970)
510. Hadni, A., *Spectrochim. Acta,* **19,** 793 (1963)
511. King, L. H., Mines Branch Research Report, R115, Dept. Mines & Technical Surveys, Ottawa (June 1963)
512. Marshall, R., Mitra, S. S. *et al.*, *J. chem. Phys.*, **43** (8), 2893 (1965)
513. Viltange, M., *Annls Chimie*, **5,** 1037 (1960)
514. DeKeyser, W. L., *Bull. Soc. fr. Ceram.*, **68,** 43 (1965)

515. Johnson, L. J., *Am. Miner.*, **49,** 556 (1964)
516. Brindley, G. W. and Choe, J. O., *Am. Miner.*, **46,** 771 (1961)
517. Duval, C. and Lecomte, J., *Bull. Soc. chim. Fr.*, **20,** 203 (1953)
518. Griffith, W. P., *Nature*, No. 5216, **224,** 264 (1969)
519. Ramdas, A. K., *Proc. Indian Acad. Sci., Sec. A*, **37** (3), 441 (1953)
520. Beck, W., *Chem. Ber.*, **95,** 341 (1962)
521. Aronson, J. R. and Emslie, A. G., *J. Geophys. Res.*, **72** (2), 687 (1967)
522. Estep, P. A., Kovach, J. J. and Karr, Jr., C., *Proc. 2nd Lunar Sci. Conf.*, **2,** 2137 (1971)
523. Estep, P. A. *et al.*, *Proc. 3rd Lunar Sci. Conf.*, **3,** 3047 (1972)
524. Perry, C. H. *et al.*, *Proc. 3rd Lunar Sci. Conf.*, **3,** 3077 (1972)
525. Ishii, M., Shimanouchi, T. and Nakahiri, M., *Inorg. chim. Acta*, **1** (3), 387 (1967)
526. Andreenko, E. D. and Plyusnina, I. I., *Vest. mosk. univ. Geol.*, **23** (4), 104 (1968)
527. Hayashi, H., *Nendo Kagaku*, **6** (2–3), 29 (1967)
528. Schubert, K. D., *3rd Int. Kalisymp. Vortr.*, **4,** 433 (1965, pub. 1967)
529. Beiseev, O. B. and Kiselev, L. M., *Geol. Razved Mestrorozhd. Tverd. Polez Iskop. Kaz.*, 319 (1968)
530. Vierne, R. and Brunel, R., *Bull. Soc. fr. Miner. Cristallogr.*, **92** (5), 409 (1969)
531. Boldyrev, A. I. and Povarennykh, A. S., *Mineralog. Sb., L'vov*, **22** (1), 11 (1968)
532. Tarte, P. and Preudhomme, J., *C.r. acad. Sci., Paris*, Ser. B., **270** (7), 474 (1970)
533. Brunel, R. and Vierne, R., *Bull. Soc. fr. Miner. Cristallogr.*, **93** (3), 328 (1970)
534. Prost, R., *Ann. Agron.*, **20** (6), 547 (1969)
535. Solntseva, L. S., *Sovrem. Metody miner., Issled*, **1,** 196 (1969)
536. Ptak, W. and Pampuch, R., *4th Conf. Clay miner. Petrol.* (Collect Pap.), p. 27 (1967, pub. 1968)
537. Bahat, D., *J. geol. Soc. Aust.*, **17** (1), 93 (1970)
538. Plyusnina, I. I. and Gribina, I. A., *Svyaz. voda dispersnykh. Sist.*, No. 1, p. 25 (1970)
539. Pampuch, R. and Ptak, W., *Pol. Akad. Nauk. Oddzial Krakowie Pr. kom. ceram., Ceram*, No. 14, p. 7 (1970)
540. Liese, H. C., *Appl. Spectrosc.*, **24** (6), 609 (1970)
541. Rocchiccioli, C., Dupuis, T. *et al.*, *J. Chim. phys.*, **67** (11–12), 2037 (1970)
542. Hunt, G. R. and Salisbury, J. W., *Mod. Geol.*, **2** (1), 23 (1971)
543. Hunt, G. R. and Salisbury, J. W., *Mod. Geol.*, **1** (4), 283 (1970)
544. Oinuma, K. and Hayashi, H., *Toyo Daigaku Kiyo Kyoyo Katei Hen* (*Shizenkagaku*), No. 9, p. 57 (1968)
545. Gillieson, A. H., *15th Colloq. Spectrosc. Int. Plenary Conf.* (*Proc.*), p. 193 (1969)
546. Balashov, V. A. *et al.*, *Tr. mosk. khim.-tekhnol Inst.*, No. 67, p. 151–4 and 154–8 (1970)
547. Kireev, F. A., *Geokhim. Sb.*, No. 4, p. 185 (1969)
548. Barsanov, G. P. and Plyusnina, I. I., *Vest. mosk. univ. Geol.*, **26** (2), 64 (1971)
549. Elderfield, H. and Chester, R., *Am. Miner.*, **56** (9–10), 1600 (1971)
550. Calvert, R. and Prost, R., *Clays Clay Miner.*, **19** (3), 187 (1971)
551. Farmer, V. C., *Soil Sci.*, **112** (1), 62 (1971)
552. Farmer, V. C. and Russell, J. D., *Trans. Faraday Soc.*, **67** (9), 2739 (1971)
553. Farmer, V. C. and Russell, J. D., *Clays Clay Miner.*, **15,** 121 (1966)
554. Farmer, V. C., Russell, J. D. and Ahlrichs, J. L., *9th Trans. Int. Congr. Soil Sci.*, **3,** 101 (1969)
555. Lindberg, J. D. and Snyder, D. G., *Am. Miner.*, **57,** 485 (1972)
556. Moore, R. K. and White, W. B., *Am. Miner.*, **56,** 54 (1971)
557. Bellamy, L. J. *et al.*, *Trans. Faraday Soc.*, **62** (12), 3435 (1971)
558. Johnson, R. A., Rogers, M. T. and Leroi, G. E., *J. chem. Phys.*, **56** (2), 789 (1972)
559. Valueva, A. A., *Rentg. miner. Syr'ya*, **7,** 196 (1970)
560. Melchakova, L. V., Plyusnina, I. I. and Sobolev, R. N., *Vest. mosk. univ. Geol.*, **26** (3), 96 (1971)
561. Tarte, P. and Thelen, J., *Spectrochim Acta*, **28A** (1), 5 (1972)
562. Preudhomme, J. and Tarte, P., *Spectrochim Acta*, **27A,** 1817 (1971)
563. Preudhomme, J. and Tarte, P., *Spectrochim Acta.*, **28A** (1), 69 (1972)
564. Yushkin, N. P. and Bushueva, E. B., *Konst. sv. Miner.*, **5,** 28 (1971)
565. Rocchiccioli-Deltcheff, C., Dupuis, T. and Wadier, C., *Compt. Rend. Acad. Sci.*, Ser. B., **273** (23), 1020 (1971)
566. Rocchiccioli-Deltcheff, C., *C.r. acad. Sci. Paris.*, Ser. B., **273** (25), 1095 (1971)
567. Shikunov, B. I. *et al.*, *Izv. akad. nauk. SSSR*, Ser. Khim, **1,** 204 (1972)
568. Miller, I. I. *et al.*, *Tr. Inst. Khim. akad. nauk. SSSR, Ural Filial.*, **23,** 109 (1971)
569. Burns, R. G. and Huggins, F. E., *Am. Miner.*, **57** (5–6), 967 (1972)
570. Baran, E. J. and Aymonino, P. J., *Z. anorg. Allg. Chem.*, **390** (1), 77 (1972)
571. Povarennykh, A. S. and Nefedov, E. I., *Geol. Zh.* (Russ. Ed.), **31** (5), 13 (1971)
572. Zorina, M. L. and Syritso, L. F., *Zh. prikl. Spektrosk*, **16** (6), 1043 (1972)

573. Cord, P. P. *et al.*, *Spectrochim. Acta*, **28A**, 1601 (1972)
574. Kharzeeva, S. E. and Serebrennikov, V. V., *Tr. Tomsk. Univ.*, **204**, 356 (1971)
375. Denning, J. H. and Ross, S. D., *Spectrochim. Acta*, **28A**, 1775 (1972)
576. Povarennykh, A. S. *et al.*, *Mineralog. Sb.*, L'vov, **25** (4) 306 (1971)
577. Petrov, K. I. *et al* , *Dokl. akad. nauk. SSSR*, **206** (4), 909 (1972)
578. Hunt, G. R., Salisbury, J. W. and Lenhoff, C. J., *Mod. Geol.*, **3** (3), 121 (1972)
579. Uklonskaya, N. T., *Zap. vses. miner. Obshch.*, **25**, 137 (1972)
580. Tarte, P. and Liegeois-Duyckaerts, M., *Spectrochim. Acta*, **28A** (11), 2029 (1972)
581. Rocchiccioli-Deltcheff, C. and Wadier, C., *Spectrochim. Acta*, **29A** (1), 93 (1973)
582. Belitskii, I. A. and Golubova, G. A., *Mater. genet. eksp. Miner.*, **7**, 310 (1972)
583. Petrov, K. I. *et al.*, *Zh. neorg. Khim.*, **18** (2), 385 (1973)
584. Petrov, K. I. *et al.*, *Zh. neorg. Khim.*, **18** (3), 833 (1973)
585. Alexanian, C., Morel, P. and Le Bouffant, L., *Bull. Soc. fr. Ceram.*, **71**, 3 (1966)
586. Carter, R. L. and Bricker, C. E., *Spectrochim. Acta*, **29A**, 253 (1973)
587. Rudnitskaya, E. S. and Lipova, I. M., *Izv. Vyssh. Ucheb. Zaved. Geol. Razved*, **15** (4), 43 (1972)
588. Ganesan, S., Maradudin, A. A. and Oitmaa, J., *Ann. Phys.* (*NY*), **56** (2), 556 (1970)
589. Runciman, W. A. and Carter, T., *Solid State Commien.*, **9** (5), 315 (1971)
590. Vergnoux, A. M. and Brunel, R., *C.r. acad. Sci.*, *Paris*, Ser. A, **265B**, (20), 1097 (1967)
591. Grimes, W. N. and Collett, A. J., *Nature* (*London*) *Phys. Sci.*, **230** (15), 158 (1971)
592. Aronson, J. R. and McLinden, H. G., *Prog. astronaut. Aeronaut.*, **18**, 291 (1966)
593. Kolesova, V. A., Izmatov, A. A. and Toropov, N. A., *Izv. akad. nauk. SSSR, Neorg. Mater.*, **5** (9), 1594 (1969)
594. Tsuyukai, N., Saito, T. *et al.*, *J. Soc. chem. Ind. Japan*, **74** (11), 2242 (1971)
595. Rodriguez, C. P., *Bull. Soc. quim. Peru*, **35** (2), 38 (1969)
596. Suknev, V. S. and Brovkin, A. A., *Zh. prikl. Spektrosk.*, **12** (2), 248 (1970)
597. Povarennykh, A. S. and Gevorkyan, S. V., *Dop. akad. nauk. Ukr. RSR*, Ser. B, **32** (2), 1[illegible]8 (1970)
598. Povarennykh, A. S., *Ber. Deut. ges. geol. Wissenschaften Riche B*, **14** (4), 291 (1970)
599. Omori, K., *Kobutsugaku Zasshi*, **9** (4), 219 (1969)
600. Levitt, S. R., Blakeslee, K. C. and Condrate, R. A., *Mem. Soc. roy. Sci. Liege. Collect 8th*, 20, 121 (1970)
601. Bogdanovich, M. P., *et al.*, *Opt. Spektrosk*, **29** (6), 1151 (1970)
602. Povarennykh, A. S. and Dustamov, V. D., *Mineralog. Sb*, *L'vov*, **23** (4), 359 (1969)
603. Solntseva, L. S. and Solntseva, B. P., *Sb. Kratk. soobshch. Mineralog. Geokhim.*, 1968*g*, **1**, 71 (1970)
604. Omori, K., *Kobutsugaku Zasshi*, **10** (1), 1 (1970)
605. Gerbaux, X. and Hadni, A., *J. Chim. phys.*, **67** (9), 1674 (1970)
606. Agiorgitis, G., *Tschermak's mineral. petrogr. Mitt.*, **13** (3–4), 273 (1969)
607. Povarennykh, A. S., *Bull. Soc. fr. Miner. Cristallogr.*, **93** (2), 224 (1970)
608. Fijal, J., *Przegl. Geol.*, **18** (1), 22 (1970)
609. Nikitina, E. I., Sotnikov, V. I. and Golubova, G. A., *Geol. Geofiz*, **4**, 114 (1968)
610. Feairheller, Jr., W. R. and Miller, Jr., J. T., *U.S. Clearinghouse Fed. Sci. Tech. Inform.*, AD 697992 (1969)
611. Likhonina, E. V., *Tr. Konf. Molodyk Isslet. Vses. Nauch-Issled. Inst. Miner Syr'ya Sekts Mineral 7th*, 94 (1969)
612. Boldyrev, A. I. and Morozova, L. I., *Mineralog. Sb.*, *L'vov*, **20** (2), 272 (1966)
613. Elderfield, H., *Instrum. News*, **21** (1), 14 (1970)
614. Povarennykh, A. S. and Dustamov, V. D., *Konst. sv. Miner.*, **4**, 3 (1970)
615. Povarennykh, A. S. and Gevorkyan, S. V., *Dopov. akad. nauk. Ukr. RSR*, Ser. B., **33** (4), 304 (1971)
616. Gevorkyan, S. V., *Konst. sv. Miner.*, **6**, 66 (1972)
617. Vlasova, E. V. and Topunova, G. A., *Tr. Mineral. Muz. akad. nauk. SSSR*, **20**, 40 (1971)
618. Aires-Barros, L., *Bol. Soc. geol. Port*, **17** (1), 51 (1969)
619. Povarennykh, A. S., *Mineralog. Sb.*, *L'vov*, **25** (3), 208 (1971)
620. Glebovskaya, E. A., *Tr. Vses. Neft. Nauch-Issled. Geolograzved. Inst.*, **294**, 76 (1971)
621. Povarennykh, A. S., *Visn. akad. nauk. Ukr. RSR*, **8**, 24 (1969)
622. Blasezak, K. and Pampuch, R., *4th Conf. Clay. Mineral. Petrology* (Collect. Pap.), p. 17 (1967, pub. 1968)
623. Kukuvskii, E. G. and Litvin, A. L., *Konst. sv. Miner.*, **4**, 81 (1970)
624. Pilipenko, A. T., Shevchenko, L. L. and Patselyuk, V. A., *Zh. prikl. Spektrosk.*, **14** (4), 638 (1971)
625. Raade, G., *Am. Miner.*, **55** (9/10), 1457 (1970)
626. Moore, R. K. and White, W. B., *Am. Miner.*, **56** (5/6), 826 (1971)
627. Ito, J. and Arem, J. E., *Am. Miner.*, **56**, 307 (1971)
628. Omori, K., *Am. Miner.*, **56**, 1607 (1971)

629. Wehrenberg, J. P., *Am. Miner.*, **56,** 1639 (1971)
630. Povarennykh, A. S. and Gevorkyan, S. V., *Mineralog. Sb., L'vov*, **24** (3), 254 (1970)
631. Friedl, W. and Krueger, H., *Opt. Acta.*, **18** (12), 912 (1971)
632. Nikonova, L. I., *Nauch. Tr. Irkutsk. Gos. Nauch-Issled. Inst. Redk. Tsvet. Metal.*, **22,** 89 (1971)
633. Larson, S. J., Pardoe, G. W. F. and Gebbie, H. A., *Am. Miner.*, **57,** 998 (1972)
634. Platonov, N. A. *et al.*, *Konst. sv. Miner.*, **6,** 62 (1972)
635. Farrell, D. M., *Can. Min. Br. invest. Rep.*, **IR 72** (1972)
636. Sakash, R. P. and Solntseva, L. S., *Zh. prikl. Spektrosk.*, **16** (4), 741 (1972)
637. Sitzia, R. P., *Re. Semin. Fac. Sci. Univ. Cagliari*, **40,** 269 (1970)
638. Lipova, I. M. and Semenov, E. I., *Izv. Vyssh. Ucheb. Zaved. Geol. Razved*, **15** (6), 50 (1972)
639. Weir, C. E., *J. Res. natn. Bur. Std.*, **70A,** 153 (1966)
640. Ross, S. D., *Spectrochim. Acta*, **28A,** 1555 (1972)
641. Zorina, M. L., *Mineralog. Geokhim.*, **4,** 58 (1972)
642. Borka, L. and Kristiansen, P. R., *Dan. Tidsskr. Farm.*, **46** (6), 130 (1972)
643. Ginsburg, I. V., *Tr. Mineral. Muz. akad. nauk. SSSR*, **21,** 179 (1972)
644. Tarasevich, Yu. I. and Sivalov, E. G., *Ukr. Khim. Zh.* (Russ. ed.), **38** (11), 1112 (1972)
645. Bilton, M. S., Gilson, T. R. and Webster, M., *Spectrochim. Acta*, **28A,** 2113 (1972)
646. Grimes, N. W., *Spectrochim. Acta*, **28A** (11), 2217 (1972)
647. Lazarev, A. N., *Spectrochim. Acta*, **29A** (1), 205 (1973)
648. Nyberg, B. and Larsson, R., *Acta. chem. Scand.*, **27** (1), 63 (1973)
649. Engel, G., *J. Solid State Chem.*, **6** (2), 286 (1973)
650. Brooker, M. H. and Bates, J. B., *Spectrochim. Acta*, **29A,** 439 (1973)
651. Paques-Ledent, M. Th. and Tarte, P., *Spectrochim. Acta*, **29A,** 1007 (1973)
652. Tarte, P., Pottier, M. J. and Procès, A. M., *Spectrochim. Acta*, **29A,** 1017 (1973)
653. Parker, T. W., *Clays Clay Miner.*, **8** (2), 135 (1969)
654. Prost, R., *C.R. acad. Sci., Paris*, Ser. D, **273** (16), 1347 (1971)
655. Prost, R., *C.R. acad. Sci., Paris*, Ser. D. **273** (17), 1467 (1971)
656. Reh, H., *Ber. dt. Ges. Geol. Wissenschaften Reihe B.*, **15,** 345 (1970)
657. Torok, I., *Agrokem. Talajtan*, **21,** 131 (1972)
658. Plyusnina, I. I. and Kapitnova, T. A., *Zh. prikl. Spektrosk*, **16** (3), 478 (1972)
659. Kuznetsova, L. G., *Mineralog. Sb., L'vov*, **25** (1), 18 (1971)
660. Povarennykh, A. S., *Geol. Zh.* (Russ ed.), **32** (5), 23 (1972)
661. Povarennykh, A. S. and Gevorkyan, S. V., *Mineralog. Sb., L'vov*, **25** (2), 100 (1971)
662. Solntseva, L. S. *et al.*, *Konst. sv. Miner.*, **6,** 30 (1972)
663. Zemann, J., *Karinthin*, **65,** 253 (1971)
664. Tarte, P. and Preudhomme, J., *Spectrochim. Acta*, **29A,** 1301 (1973)
665. Akhmanova, M. V. and Alekhina, L. G., *Probl. izomor. Zameshch. At. Krist*, p. 243 (1971)
666. Ericksen, G. E. and Mrose, M. E., *Am. Miner.*, **55,** 1500 (1970)
667. Mrose, M. E., Fahey, J. J. and Ericksen, G. E., *Am. Miner.*, **55,** 1518 (1970)
668. Tarasevich, Yu. I., *Kolloid. Zh.*, **33** (1), 135 (1971)
669. Adler, H. H., *Econ. Geol.*, **58** (4), 558 (1963)
670. Fleischer, M., *Glossary of Mineral Species*, Bowie, Maryland. Mineralogical Records Inc. (1972).
671. Hey, M. H., *Chemical Index of Minerals*, London, British Museum (Nat. Hist.) (1962); and Appendix (1963)
672. Grim, R. E. and Kulbicki, G., *Am. Miner.*, **46,** 1329 (1961)
673. Florinskaya, V. A. and Pechenkina, R. S., *Glass Industry*, **39,** 27, 93 and 151 (1958)
674. Hunt, J. M. and Turner, D. S., 'Determination of Mineral Constituents of Rocks' (quartz and carbonates mainly), *Analyt. Chem.*, **25** (11), 1169 (1953)
675. Lyon, R. J. P. and Tuddenham, W. M., 'Determination of Kaolin Group Minerals' (halloysite, kaolinite and dickite), *Nature*, **185,** 835 (1960)
676. Moenke, H. and Moenke-Blankenberg, L., 'Quantitative Infrared Spectrophotometric Determination of Dolomite and Calcite Content of Sedimentary Carbonates', *Janaer Jahrbuk*, Pt II, p. 396 (1960)
677. Fisher, R. B. and Ring, C. E., 'Quantitative Infrared Analysis of Apatite Mixtures' (fluorapatite in hydroxyapatite), *Analyt. Chem.*, **29,** 431 (1957)
678. Scholze, H., 'Quantitative Investigation of Water in Silicates' (opal, glass? and muscovite), *Fortschr. Miner.*, **38,** 122 (1960)
679. Corbridge, D. E. C. and Lowe, E. J., 'Quantitative Infrared Analysis of Condensed Phosphates', *Analyt. Chem.*, **27,** 1383 (1955)
680. Lyon, R. J. P., Tuddenham, W. M. and Thompson, C. S., 'Quantitative Mineralogy in Thirty Minutes', *Econ. Geol.* **54** (6), 1047 (1959)
681. Tarte, P. (the distinction by infrared spectrometry between mullite and sillimanite), *Silic. Ind.*, **24,** 7 (1959)

682. Chester, R. and Elderfield, H., 'Total Carbonate in Marine Carbonate Sediments', *Chem. Geol.*, **1** (4), 277 (1966)
683. Dune, R. A. and Rosevear, P. E. (estimation of minerals in beach sands), Invest. Rep. 85 of the Div. of Mineral Chem., S.I.R.O., Sydney Laboratory (July 1970)
684. Bartl., M. (determination of water of crystallisation in gypsum in the presence of clay minerals), *Analyst*, **97**, 854 (1972)
685. Chasan, D. E. and Norwitz, G. (determination of barium, strontium, sodium and potassium sulphates), *Talanta*, **18**, 499 (1971)

Subject and Chemical Index

Mineral Index